OCEAN LIFE
BENEATH THE CRYSTAL SEAS

OCEAN LIFE
BENEATH THE CRYSTAL SEAS

THE IMAGE BANK®

ISBN 0-941267-15-6

Manufactured in Spain

Producer : Ted Smart
Author : Vic Cox
Book Design : Sara Cooper
Photo Research : Ed Douglas
Production Assistant : Seni Glaister

THE SOURCE

Many of the world's creation myths trace the wellspring of life to water. Marine sciences may not back these myths in every detail, yet they do provide overwhelming evidence for life originating in seawater billions of years ago.

Since science and myths often collide in the modern world, it is comforting that when it comes to the ocean there is room for compatibility. Perhaps that is no surprise for a medium that covers better than 70 percent of Earth's surface, comprises 98 percent of its volume, and reduces the continental land masses to islands. The ocean even pushes the land around, albeit very slowly. Continents rest on crustal plates constantly in motion due to molten rock squeezing from seabed fissures miles below the surface.

Even the greater part of oxygen in the atmosphere derives from uncountable microscopic plants, called phytoplankton, in the sun-lit top layer of the world ocean. These plants absorb carbon dioxide and release oxygen in a process known as photosynthesis. Life on land is permanently linked to life in the sea, no matter how different the living conditions seem to be.

LIGHT! ACTION!

Light can penetrate very clear water to about 350 feet and still power photosynthesis. It is much less in murky water. These few hundred feet are called the surface zone or layer. This zone contains only 2 percent of the volume of the world ocean, but perhaps as much as 90 percent of its living organisms.

Floating on or near the surface are plants and animals collectively known as plankton. Plankters take many shapes and range in size from microscopic to jellyfish three or more feet in diameter, but there are two primary kinds: the plants (phytoplankton) and the animals (zooplankton). Absorbing energy from the sun, the surface phytoplankton converts nutrients into sugars and starches. Zooplankton, which moves up and down below the surface, consumes the energy stored in plants and any larva within reach. Other carnivorous organisms eat the tiny animals and, in turn, are eaten by larger carnivores up to tunas and sharks.

Though most zooplankters are poor swimmers, they easily alter their buoyancy. Daily they commute between the surface and oxygen-rich colder water hundreds of feet down, arranging themselves at preferred levels of temperature and pressure. Major concentrations have been found at depths of 500 and 5,000 feet; some communities have been located as deep as six miles. It is doubtful these abyssal animals vertically migrate, but others in the water column ascend from as much as 1,600 feet as the light fades. It is the single largest daily movement of animals anywhere on the planet, including commuters on the Los Angeles freeways.

In the ocean, individuals learn to go with the flow early in life, and that flow has direction. The earth's rotation imparts a spin to water movement so that north of the Equator surface currents move clockwise; to the south, counterclockwise. This produces nearly closed, circular patterns called gyres. Surface currents are driven by seasonably variable winds and the heating and cooling of the waters, which is influenced by polar cold water currents crossing the seabed toward the Equator.

Some surface currents, such as the Gulf Stream of the North Atlantic, are like a system of converging rivers that speed up and slow down at different points. Off Canada the equatorially warmed Gulf Stream mixes with the cold, Europe-bound North Atlantic Current. The Gulf Stream's warm waters temper the frigid impact of Arctic air masses. Thousands of years ago this phenomenon helped make northern Europe habitable for humans.

THE SARGASSO SEA

Aside from mixing equatorial waters with cold waters, surface currents may create peculiar localized conditions. The best-known example is the Sargasso Sea. Propelled by the Gulf Stream, two other major currents and the cold Atlantic, the warm Saragasso rotates slowly east of Bermuda like a southern gear wheel in a conveyor belt circling the North Atlantic. Little rainfall and lots of evaporation make this gigantic eddy highly saline. There is no mixing with deep, cold water-borne nutrients, so the Sargasso is clear and cobalt blue—except for the tawny islands of drifting weed that give the sea its name.

Sargasso weed (*Sargassum natans* and *S.fluitans*) is the open ocean's largest seaweed. It has no root system, taking nutrients directly from seawater, and floats on clusters of grape-sized gas bladders in clumps of brown algae. Its ancestors may have been coastal dwelling rockweeds, but *Sargassum* has evolved into a drifter scattered across a stretch of the Caribbean Sea nearly as wide as the United States.

One transient fish with a surprising link to the Sargasso is the common eel (*Anguilla anguilla*). For millennia, naturalists have known that adult freshwater eels leave Europe's rivers every autumn for the sea—and never return. However, young eels, looking like a cross between a finfish and a mature eel, do swim up these rivers from the Atlantic. Not until this century was most of the eel's life cycle pieced together and the Saragasso Sea identified as its spawning ground.

Mature North American common eels (*A. rostrata*) join their European cousins by the millions in the autumnal migration to the southwestern Atlantic. Major physical changes, which start in their freshwater homes, unroll en route. The eel's sex organs mature; its color turns metallic silver with a purple-black stripe down its back. Lateral sensing lines along the sides widen and eyes double in size. Where the retina was purple to better collect the blue-green light of fresh water, it changes to gold, which easily absorbs the blue light of the Saragasso.

Apparently traveling at great depths, the eels swim 1,000 miles or more to the Sargasso where females release between 15 and 20 million eggs each. Though the exact spawning sites have yet to be discovered, pea-sized eel eggs have been found from 650 to 1,800 feet down. Since no adult eels of either species have ever been caught here, it is suspected that, like the marine conger eel (*Conger oceanicus*), the common eels disintegrate after egg laying.

What stretches the imagination is the migration of their offspring, which start life looking like flatfish made of cloudy glass. Only a quarter-inch long, the *leptocephali*, as eel larva are known, drift with other plankters caught in the Sargasso's slow circle until they catch the Gulf Stream northbound. Governed by different clocks, the juveniles swim the Atlantic for one year if they are North American eels and about two and a half years if they are European eels before entering their respective rivers.

NATURE'S HALFWAY HOUSE

Triggered by the pull of the moon and, to a lesser extent, the sun on the world ocean, tidal action sculpts a halfway house where land and sea join in dynamic coexistence. Flooding and ebbing creates the tidal marsh, or estuary if it is a river's mouth, and companion mud flats.

Wide and frequent swings in temperature and saltiness characterize these wetlands, yet they are among the most productive zones along the world's shores. They are termed nature's nurseries because, for at least one stage in their development, an estimated 90 percent of the important coastal commercial and game fish use them. Wetlands are dominated by cordgrass species in temperate climates and mangrove thickets in the tropics. Large reefs of mussels (*Mytilus edulis*) grow in estuarine mud flats, anchored by wrapping tough byssus threads around anything solid, and filtering food from the tides and organic runoff.

Among the unusual marine animals of this zone is the mudskipper (*Periophthalmus*), which spends more time out of water than in it. The 5- to 12-inch-long tropical fish lives in soft mud with only bulbous eyes showing. It has no lungs but can fill its gill chambers with air and water, allowing it to remain out of water for some time. This frees it to climb the roots of mangrove trees after insects.

Many air-breathing reptiles, like sea turtles and the salt water crocodile, are born in the wetlands. Though their long lives are mostly spent at sea, turtles bury their eggs above the high-tide level on only certain stretches of sandy beaches. While most brood nests are sited on both sides of the Equator, green turtles (*Chelonia mydas*) are found in the eastern Mediterranean and loggerheads (*Caretta caretta*) use sites in south Florida and southern Africa. The leatherback (*Dermochelys coriacea*) is the largest and most ocean-going of the seven marine turtle species. It often weighs over 1,000 pounds and exceeds 6 feet in length.

Salt water crocodiles (*Crocodylus porosus*) are egg layers, too, but they build nesting mounds on banks of marshes and estuaries along the shores of southeast Asia and northern Australia. Unlike turtles, these and other crocodiles stay close to their mounds until the eggs hatch. A full-grown salt water croc may reach 20 or more feet, making it the largest member of the crocodile family.

Mountains crumble and miles of rock are alternately stripped and coated with sand by tidal power, and yet life thrives in the combat zone between land and sea. Depending on whether it is a sand- or rock-dominated environment, crabs, periwinkles or littoral snails, and pill bugs share the mostly dry, high-tide splash zone with algae and black lichens. Avoiding all but the spray, *Littorina* clings to rocks three feet above the high-tide line, breathing air and wetting its gills infrequently. One European species of periwinkle has no gills and drowns if kept submerged. Tropical species have been seen on grasses, bushes and trees bordering the sea.

Grunion (*Leuresthes tenuis*) are a tidal zone fish that comes ashore between March and August to lay eggs in certain California beaches. Their egg-laying is timed to the highest tides following the full or new moons. The 6-inch-long females bury their lower bodies in the sand and the males curl around them, dumping large amounts of sperm. While the process takes about 30 seconds, some females may be out of water several minutes.

Laying eggs on beaches is an ancient rite, as a living fossil illustrates. The horseshoe crab (*Limulus polyphemus*) is a bottom-dwelling animal that hunts during the day by burrowing into sand bars in search of small animals. Growing as long as 20 inches, the horseshoe crab looks the same as it did 250 million years ago. Its ancestral line extends back 400 million years.

The horseshoe is found along the coasts of New England and the Gulf of Mexico in the U.S., and around Japan and southern India—very different regions of the world ocean. In early summer, the female, which is much larger than the male, crawls onto the beach to deposit her eggs. She often drags along a male or two who has hitched a ride on her taillike spike. After she has scraped out a shallow pit and laid 200 or 300 small round eggs, the male sits on her back and dumps sperm on the eggs.

Contrary to its popular name, the animal is a distant relative of the spider, not the crab. A horseshoe crab has two sets of eyes mounted on a rounded shield, and walks upon the seabed with five pairs of short, jointed legs. Since the animal's mouth is centered on its underside, the spines at the base of the legs also serve as food grinders. As the legs move, smaller pieces transfer to the next pair of "jaws." Eventually, the food ends up in the mouth. The leg spine arrangement gives unexpected antiquity to the act of eating on the run.

ROCKY COASTS HUGGERS

Life in the estuaries, mud flats and sandy beaches has to survive quick changes in salinity, temperature, and frequent exposure out of water. Things are not much more stable in tide pools and surf zones. Here wave action places a high premium on being able to take a constant pounding. No creature does this better than the acorn barnacle (*Balanus*), the small white crustacean that builds its hard teepee on just about any rocky shore.

Like an array of miniature volcanoes, bands of acorn barnacles mark the reach of high tides along rocky coasts. Standing on their heads, they skim food with curled, featherlike legs from the swirling water. Most barnacle species are self-fertilizing and release larva when currents will sweep them out to the plankton meadows.

After several weeks of free-swimming the larva is physically ready to settle. Its timing is influenced by factors such as surface texture, light, currents, depth, and especially whether other members of its species have already homesteaded the neighborhood. Each surface contacted is chemically tasted by a special appendage on the larva that can analyze a substance's molecular composition. If the surface passes muster, the barnacle spat makes room to grow—other spat may be bumped as the newcomer seeks a clear spot—and glues its head down with an internal cement that is chemically similar to its skin.

In the water world, any solid foundation or substrate is sought by plants and animals for which shelter is top priority. Besides ship hulls and rocks, barnacles grow on stationary bivalves and snails; they also are found on sea turtles, horseshoe crabs, and even on some whales. Encrusting sponges vie with sea squirts for space on a slender bryozoan, a moss animal that has twiglike branches. Even anemone-crowded tide pools on a rocky coast reflect the struggle for living space inside a tidal flow of nutrients.

Sea flowers, as anemones are sometimes called, do not just hunker down against the surf. While firmly attached by foot disk to rocks and boulders, they have muscular stalks that flex with the force of waves and currents. Hollow tentacles coated with stinging capsules, called nematocysts, fringe mouths usually an inch or two wide but which grow up to three feet across among tropical species. At low tide when they can no longer feed, the animals draw their colorful arms into stalks that are often camouflaged with sand and bits of shell.

Anemones are usually the predators, firing banks of nematocysts into fish that come too close and stuffing the stunned animal into their mouths. While sensitive to physical and chemical contact, vibrations from a swimming prey have recently been shown to trigger nematocysts. But there are other, even more vibrantly decorated animals that feed on anemone tentacles with impunity. These are the sea slugs, or the nudibranchs, which means "naked gills."

BEASTLY BEAUTIES

Nudibranchs, which may be arrayed in soft pinks, electric blues, shocking oranges and even virginal whites, have discarded the cumbersome shells that protect most ocean-going mollusks. In place of the mobile fortress many sea slugs have evolved a form of chemical defense.

Out of water these gaily painted animals emit a penetrating, fruity odor. Fish apparently are susceptible to the sea slug's perfume, for these nudibranchs have few natural predators. In laboratory experiments, unwary fish engulf adult sea slugs, but quickly spit them out.

There are several branches on the nudibranch family tree, but two major ones are the eolids and the dorids. One of the loveliest sea slugs, the opalescent nudibranch (*Hermissenda crassicornis*), is a classic eolid. It has a well-defined head, which resembles that of its relative the common garden snail, and packs a row of fleshy spikes on its slender back.

Eolids breathe through these spikes, or cerata, which sway in the current and seem quite delicate. Again the nudibranch's appearance is deceptive. Their cerata may be chock full of nematocysts gathered by munching on soft-bodied animals like anemones and hydroids or, in tropical waters, coral polyps.

Dorids breathe through a ruffle of skin around the anus and have wide, flat bodies. A typical dorid is the lemon nudibranch (*Anisodoris nobilis*), which is one of the largest sea slugs along the California coast. Some of these yellow animals sprinkled with orange polka dots grow as long as 8 inches. Larger Australian versions of this species have been reported.

Evidence suggests the nudibranch's color depends in large part on the zone of water in which it lives. The lighter, brighter-colored sea slugs tend to inhabit the shallow waters while their darker, duller relatives tend to live deep in the ocean. Dark brown nudibranchs have been dredged up from as deep as three miles, but little is known about them.

THE KELP FOREST

Seaweeds are marine algae which are distributed worldwide according to water depth, temperature, and light penetration. Green algae, like eel grass, shares shallow waters with brown algae, such as kelp. The brown plants are called that because other pigments cover their green chlorophyll. Red algae is located in deeper waters where light rays are the weakest. All are important sources of food and shelter for entire marine communities and many are useful to humans.

Algae were burned for soda and potassium salts in Europe at least as far back as the seventeenth century. The ashes, which contained the salts, were originally called kelp. Now the word refers to the brown algal seaweeds, like giant kelp (*Macrocystis pyrifera*), bull kelp (*Nereocystis*), feather-boa kelp (*Egregia*), and the smaller species known as *Laminaria*.

Unlike most land plants, kelps have no root system for food gathering. They absorb nutrients throughout the whole supple structure: blades, stalk (or stipe) and holdfast, a tangle of tough, fibrous strands at the plant's base that attach to rocky outcroppings. These algae cannot develop on mud or sand bottoms. The major exception is the drifting gulfweed of the Sargasso Sea.

At a rate of up to 18 inches a day, giant kelp is one of the world's fastest-growing plants. This growth, and the commercially extracted alginic acid produced by kelp, make it a valuable and renewable resource. But to many of the myriad animal larva wandering through the coastal upper waters, kelp forests simply mean survival. It is estimated that the average bed of giant kelp provides 14 times the encrusting surface than would be available on the bottom alone.

Seaweeds are a happy hunting ground for predators and prey alike. Kelp snails graze on floats while nudibranchs inch across blades. A yellow kelp crab climbs a stipe while a hydroid spreads branchlike arms from its base on the holdfast. On the seabed, purple and red sea urchins munch on pieces of *Macrocystis* that have drifted to the bottom. Knobby skinned seastars persistently probe the terrain for shellfish meals. Flashing schools of fish, some of which feed on kelp shoots and others that graze on the grazers, are common visitors in this dappled realm. Individual residents, such as kelp bass and tree fish, sport colors and scale patterns that blend in with their surroundings.

Atop the forest floats the sea otter, a small marine mammal with a large appetite for urchins, shellfish, and many other kelp residents. The otter prunes the grazers, keeping their numbers manageable. It relies on a thick pelt for insulation rather than a blubber layer like that encasing seals and whales. The otter also wraps itself in giant kelp's rubbery fronds when it wants to nap.

Giant kelp depends on abundant seed to multiply. Each spring special fruiting blades release millions of spores. Propelled by two tiny tails, the spores work their way to the bottom and develop into microscopic male and female plants. The sperm and eggs of these plants must combine to grow new giants.

RAPPING URCHINS

At first glance the purple and red pin cushions dug into the littoral's rocky ridges look like a subsea version of "The Untouchables." Sea urchins, formidable as they seem, are far from invulnerable. Sheepshead fish and wolf eels crunch them; sea otters crack them like eggs, and sea birds grab them at low tide for one-way rides, smashing urchin shells on rocks and highways.

Chief among the sea urchin predators is the twenty-armed sunflower seastar (*Pycnopodia helianthoides*). Seastars, or starfish, are the urchins' cousins, but their relationship is strictly that of the diner to its meal. As the sunflower slowly clambers among the giant red urchins (*Strongylocentrotus franciscanus*) it stirs a ripple of defensive action. Too slow to flee, the nearest urchins swing their spikes on ball-and-socket joints. A phalanx of purple and red spears confronts the tiny tube feet extending from the seastar's arms. Usually, this is not enough to stop a sun star.

The sunflower mounts its prey and turns its stomach inside out, covering the urchin. Strong digestive juices dissolve the muscles anchoring the spines to the shell. Eventually, the sunflower overturns its victim and forces its flexible stomach through the urchin's mouth.

With the thick, five-armed starfish the urchin has a fighting chance. In addition to spikes, the urchin deploys its own tube feet and the stalked pincers tipped with poison glands that are called pedicellariae. Grappling with the seastar in foot-to-foot combat, the urchin's tube feet and pincers are torn away. But sometimes a well-dug foxhole and the pain of repeated stings save the urchin. The tropical urchin *Asthenosoma* has the added defense of poisonous spines.

Defense is not the spines' only function. The urchin can change location or, with help from a rasplike mouth, bore into soft rock by moving the spines in rippling waves. Some species also move with the help of tube feet and a suction system that drags the animal forward. Spearing kelp blades, spines transport food to the underside of the shell where five sharp teeth rasp algae into bits.

In the mid-1970s, researchers discovered that adult urchins play vital roles in the survival of not only their young but also juveniles of other animals. Young shrimps, abalones, crabs, asteroids, snails, and fishes routinely seek shelter under the urchins' spines, concluded a study by Scripps Institution of Oceanography scientists in La Jolla, California. If the juveniles shared the urchins' fondness for brown algae, they benefitted from free meals, too.

The concept of sea urchins as spiney nursemaids came as a shock to those who saw them as invaders laying waste to valuable kelp forests. While *Macrocystis* is plentiful south of the Equator, beds are restricted to a few areas along North America's Pacific coast. California alone once had around 100 square miles of kelp forests, mostly in the southern half of the state. But urchins were only part of the reason behind California's loss of 90 percent of its kelp beds in the last several decades.

Sea urchins normally scavenge kelp, but they will forage if their population is dense and predators scarce. So urchin swarms gnawed through holdfasts, allowing fronds to drift away. Reduced kelp supplies usually would have limited their activity or thinned them to whatever was the area's new carrying capacity. But humans added new factors.

In California's case, decades after otter hunting ceased the human factor took the form of sewage pipes. This provided urchins with steady nourishment, and they multiplied. Combined with an unexplained warming of the coastal water, sewage-fed hordes ravaged seaweed forests and damaged a multimillion-dollar kelp-based chemical industry.

After years of research and undersea campaigns to annihilate urchin armies, the 1970s saw a comeback for giant kelp. People replanted kelp beds after the areas were cleared of urchins and, in the late 1980s, otters were experimentally reintroduced at San Nicholas Island off southern California. More importantly, a commercial fishery arose from the urchins' shattered shells. In Mediterranean countries and Japan the sex organs are considered delicacies and cost more per pound than abalone. Eventually, American divers began exporting some species of sea urchin.

MILD-MANNERED MANATEES

Another coastal plant-eater is the West Indian manatee (*Trichechus manatus*). It has a face only a mother or another manatee could love: a blunt snout with nostrils mounted on a boxlike head two sizes too small for an oval-shaped body 8 to 10 feet long. No beauty contest winner, it is a remarkably gentle creature without any known natural enemies.

The sea cow, as it is commonly called in its Florida coastal home range, is a marine mammal that lives wholly off plants in salt or fresh waters. It has broad front flippers that bend at the wrist to help shove leaves to its muzzle. Teats under the flippers allow its calves to nurse.

Florida sea cows leave their range along the Atlantic and Gulf of Mexico coasts in winter to seek warmer havens in rivers and estuaries. Though experts disagree on their exact threshold, it is believed that manatees cannot endure temperatures below 60 or 65 degrees. (*Note:* All temperatures are in Fahrenheit.) Pneumonia is a frequent result, often ending in death. Florida's many rivers, especially those with warm effluents from power plants, have proven irresistible winter retreats.

Increasingly, the banks of these bays and rivers are sprouting marinas, resorts, and residential developments for the state's burgeoning human population. Motor boats roar through channels where placid manatees graze. It is not unusual for whirling props to slice across the backs of dark, subsurface feeders. Encounters with humans, many of which are unintentional, are the chief cause of death among the manatee.

With a population of less than 1,200, manatees are one of the most endangered of United States marine mammals. They were hunted by native Americans and European settlers for meat, oil and skin, but never at the commercial level of fur seals or sea otters. State and federal protection in this century has helped reduce deaths, but poaching continues to add to a mortality rate most experts consider way too high—especially when gestation is estimated to be between 385 and 400 days.

Manatees reside in the Caribbean and along Central and South American Atlantic coastlines. A subspecies lives in the rivers and along the coast of West Africa. They concentrate in river deltas, though their numbers are sparse everywhere. A similar-appearing relative, the dugong (*Dugong dugon*), swims the tropical shallows of the Indian Ocean and Australia's north coast.

CORAL REEF HAVENS

It grates against the romantic image of white sand tropical atolls fringed with coral reefs, but the animal that makes it all possible is a tentacled killer. The hard coral polyp looks so much like a sea anemone atop a limy cup, known as a corallite, that it is no surprise they are related coelenterates. Jellyfish are members of this same animal group, which is characterized by tentacles usually loaded with nematocysts. The polyp stores its soft tentacles inside the corallite during the day when predators are most active; at night they sprout deadly petals and feed.

Corals may be soft or stony, solitary individuals or colonies of physically interconnected animals. They have adapted to living at various depths and temperatures in many areas of the ocean, but share a similar physical system and develop from a founder polyp. The individual animal may remain separate, such as the cold-water Devonshire Cup coral (*Caryophyllia smithii*), or rapidly subdivide into a network of organically linked polyps, as do the reef corals.

The great variety of hard coral colonies' shapes and sizes is suggested by some common names: staghorn coral, brain coral, mushroom coral, lettuce coral. These formations develop according to how individual polyps build corallite cups. Each polyp may be as little as an eighth of an inch wide at its mouth or as much as 39 inches across, depending on the species. All of the colony are connected by a mantle of living tissue so that when one group of polyps gathers food into their digestive tracts, the collective whole is fed.

Reef-building corals grow only in tropical waters. To flourish, reef-builders require clear water warmed to more than 64 degrees, ample sunlight and no silt. Many species of polyp harbor algae, called zooxanthellae, in their tissues that are vital to the host. Through photosynthesis the plants contribute food and help the polyps extract calcium carbonate from seawater. The minerals are used to construct the limestone cups. Though zooxanthellae are not universal among reef-building corals, studies indicate that all of the fastest-growing hard corals have the microscopic plants.

Sea fans and sea whips are slender, branchlike gorgonian corals that shelter polyp colonies inside tubes of horny rather than stony material. This gives the animals flexibility, allowing them to bend with the currents. They are abundant in warm water, but many *Gorgonia* are at home in temperate and cold waters, too.

Over millions of years, layer upon layer of corallite and coralline algae created massive atolls atop sea mounts, many of which are dead volcanoes. Eniwetok Atoll, one of the Marshall Islands in the Pacific, has coral rock 4,875 feet thick. Seismic readings of Bikini Atoll found 6,500 feet of coral material atop bed rock; the living reef consisted of just the top 180 feet. Such solid bases were deemed ideal for the nuclear bomb testing conducted by the U.S. on these islands after World War II.

In the nutrient-poor shallows, coral reefs provide havens for diversity. Their colorful residents form a complex ecosystem that is much studied but not fully understood. Australia's Great Barrier Reef indicates how large that system can get. It runs an estimated 1,600 miles along the island continent's northern coast and averages 95 miles in width. It is home to more than 3,000 animal species, some of which seem virtually defenseless against predators.

Even hard corals have no protection against specialized feeders like the parrotfish (*Scarus*), which can munch through corallite with beaklike teeth to reach polyps. The crown-of-thorns starfish (*Acanthaster planci*) ignores nematocysts as it cuts swaths through polyps, pumping their cups full of digestive juices and devouring the broth.

Soft corals, which have no cup and form no part of the permanent structure, nonetheless cover significant tracts of a reef. A single soft coral colony may grow to more than three feet in diameter. Sometimes called tree corals, certain soft corals make needle-sharp limestone thorns, which could give a fish pause before taking another bite. But most species seem to rely on a chemical defense.

These soft corals generate large quantities of terpenes, a chemical common to many trees. Stony corals do not. Indeed, when soft corals attempt to overgrow their limy cousins they release a barrage of terpenes, which kills the hard corals.

In nature, as in human warfare, every measure has a counter-measure. The egg cowrie (*Ovula ovum*) dines on soft corals, somehow transforming terpenes into a nontoxic substance in its digestive gland or liver. Other snails with beautifully decorated shells, the triton (*Charonia*) and the helmet shell (*Cassis*), hunt the crown-of-thorns starfish. They are important natural controls because each adult starfish releases up to thirty million eggs into the plankton. Normally, the triton and helmet shell snails are common throughout the shallows of the Indo-Pacific, but they are also widely collected for the international shell shop trade. When *Acanthaster* predators were needed, these snails were absent.

Among the many other reef animals are multicolored forests of tube worms, which like to populate coral heads, swaying banks of tunicates or sea squirts, both shelled snails and unshelled sea slugs, starfish, sea urchins and sea cucumbers. Sponges at various levels of the reef range from giant vase and elephant ear species, to tubular kinds resembling pipe organs, to thin encrusting mats like the breadcrumb sponge.

Inside the main channels of large sponges are other animals, like shrimp or small fish. Pistol shrimp and mantis prawns seek sandy bottoms, and crabs hide under coral boulders and among the seaweeds. No niche goes unfilled for long on a reef.

SCORPIONFISH

One species of coral reef scorpionfish, *Pterois volitans*, is a particularly striking fish. Commonly known as lionfish, zebrafish or turkeyfish, *Pterois* is a compact, 3-inch terror of stripes and plumed spines packing potent venom. The lionfish has been seen attacking prey and jabbing them.

Its aggressive behavior is unusual among members of the worldwide scorpionfish family, which include the temperate- and polar-water rockfish and sculpin as well as the lethal tropical stonefish. There are hundreds of species of *Scorpaenidae* so generalizations often have exceptions.

Rockfish (*Sebastes*) species can grow as long as 27 inches but are more commonly half this size. They have tasty flesh and are a popular Pacific sport fish. Since World War II, they have become a major international fishery with a catch running around 350,000 metric tons annually.

All rockfish have internal fertilization and bear live young. A 20-inch olive rockfish (*S. serrannoides*) may spawn as many as 650,000 larval offspring; bocaccio (*S. paucispinis*) and vermilion rockfish (*S. miniatus*) a few inches longer have been estimated to produce up to 1.6 million young. Both of the latter species' young feed on plankton in the open ocean and have been caught as far as 350 miles offshore. *Sebastes* do not do well in the tropics, but other species of the scorpionfish clan thrive in warm waters where they have developed potent poisons.

The tropical stonefish (*Synanceja horrida*) is a misshapen shallow-water fish that waits, partially buried in sand and mud or under seaweed, for food to come within pouncing distance. Its mottled colors and fleshy lumps break up any recognizable outline so it is often mistaken for a mound of mud or debris. The 2- to 3-foot-long stonefish lives in tide pools along the coasts of India, China, the Philippine Islands, and Australia.

Nicknamed goblinfish and warty-ghoul, this fish discharges 5 to 10 milligrams of venom from each spine. Results range from muscular paralysis to shock, and from difficulty in breathing to cardiac arrest. The effects are always accompanied by intense pain. To top it off, the stonefish can inflict a deadly sting after considerable time out of water, as some victims have learned to their dismay.

NAUTILUS: THE ORIGINAL ANCIENT MARINER

Nautiluses (*Nautilus*) are a family that can be traced back 500 million years. Squid and octopus are related, and may have roots that stretch nearly that far back. The distinctive spiral shell is a change from long, tusklike shells that sometimes grew to 30 feet. Like the coelacanth, a living fossil fish thought extinct millions of years ago, the nautilus survived extinction but has left a 5-million-year gap in its fossil record. That was how long ago the last nautiluslike animals lived off North American and European shores.

There are believed to be about half a dozen modern species and the adult shells now range from 4 inches to 12 inches across. Like the octopus and squid, they propel themselves by squeezing jets of water through a siphon and catch prey with tentacles.

Nautiluses take fifteen to twenty years to reach sexual maturity, which is why there is great concern that they may have been overfished. Their unusual method of reproduction also contributes to slow population growth. Every time they mate, the male sex organ breaks off and stays implanted in the female. The male then has to grow another sex organ before mating again.

Nautilus eggs are among the largest created by any soft-bodied animal, reaching 1.5 inches long. Few are produced by the female. Daily temperature changes have recently been discovered to be crucial to producing fertilized eggs. Baby nautiluses hatch with a shell already an inch long and are immediately on their own. Internal buoyancy control enables them to ride currents over a wide range.

Most nautiluses today are found in the western Pacific off coral reefs. They are also known from several areas in the Indian Ocean and their shells have been discovered on Kenya's beaches. However, these mollusks are great drifters. A living nautilus has been known to drift from its home in the Philippines thousands of miles to Japan.

While adult nautiluses can travel from as deep as 1,800 feet, allowing the animals to avoid daytime predators, their survival tactics are not well known. But they are scarce, and no one knows if they are being pushed to extinction.

SHARING THE TABLE

Not all sea animals are always out to eat their neighbors. The mutually beneficial bond between some coral and plant species has been noted, but there are examples of cooperation between predator and potential prey, too. Coastal and coral reef residents often have areas where small animals clean large animals of parasites, flaking scales, and encrusting organisms that may harm the client but nourish the cleaner. These are known as cleaning stations.

Cleaner wrasses (*Labroides*), a small striped fish, use their striking color design and movements to signal their desire to clean very large clients, such as groupers. The similarly patterned sharknosed goby (*Gobiosoma evelynae*) is another fearless cleaner fish: It even swims inside the client's mouth and out its gills.

Female prawns of the genus *Periclyemes* live among anemones of the *Stoichactis* group. The prawn keeps the anemone clean of detritus and the anemone protects the crustacean, particularly around spawning time. This same family of anemones has a special relationship with the white-banded clownfish (*Amphiprion*) that allows this potential meal to rest unharmed among the tentacles.

Adults of all 26 species of Indo-Pacific clownfishes are only found among sea anemones, according to marine biologist Daphne Fautin, a curator with the California Academy of Sciences in San Francisco. The clownfish, once it transforms from its free-floating larva stage, dwells among an anemone's tentacles for the rest of its life. It is a poor swimmer, seldom traveling more than a few feet from its protector, and is quick to dart back when a wrasse or other predatory fish nears.

Unlike some biologists, Fautin thinks the clownfish also helps the anemone. She cites the butterflyfish's (*Forcipiger*) appetite for anemones and corals. In one experiment, anemone species on Australia's Great Barrier Reef and in Papua New Guinea were eaten by butterflyfishes after Fautin removed their clownfishes. She notes that clownfishes bare their teeth and "chatter," or make other threatening displays, when butterflyfish approach their anemones.

As to why the clownfish can swim among the tentacles without being stung, Fautin and other researchers say that the mucus-covered fish most likely secretes something to inhibit the anemone's attack. However, a prolonged separation of the former partners results in stings if the fish returns. When this happens the clownfish reintroduces itself to the anemone with quick, passing touches. Apparently, the action transfers mucus from the anemone to the fish and the coating helps reestablish the commensal arrangement.

WHALES AND DOLPHINS

Although human activities like hunting and oil spills have destroyed large numbers of sea animals, no group has suffered a more drastic decline than the great whales. In this category are whales whose sizes range from the 30-foot minkes (*Balaenoptera acutorostata*) to the 90-foot blues (*Balaenoptera musculus*). They include 10 species of baleen, or mustached, animals and one toothed whale—the sperm. Population estimates are always tricky, and experts can reasonably disagree, but the current guesstimate is that less than 3 million great whales swim the world ocean today, 90 percent of which are sperms and minkes.

It is hard to determine the accuracy of such numbers. The International Whaling Commission's (IWC) scientific committee down-graded its estimate of blue whales in the Antarctic from several thousands to less than 500 in its 1989 report. Such wide fluctuations are unusual, but they illustrate the unsteady state of whale population estimates. Any figures for the prewhaling era are of necessity highly speculative. Commercial whalers targeted the largest whales they could kill, particularly those that fed on tiny crustaceans and small fish in the nutrient-rich waters of the polar seas.

These mustached whales comb tiny organisms from the water with baleen, a series of stiff but flexible vertical plates with a fringed edge. The plates hang from the interior of the upper jaw in closely packed rows. When the whale's tongue forces water out of its mouth, the baleen, which is made of the same material as human fingernails, captures hundreds of pounds of food.

With commercial whaling suspended until 1991, the prime cause of cetacean mortality today is the incidental killing of dolphin and porpoise in the nets of various fisheries. In what may be a low estimate of the incidental kill, the IWC calculates that around 100,000 small cetaceans die each year, most of them unwanted victims in sets for various fish. Some, however, are caught because of the valuable fish with which they swim.

Deep-water spinner (*Stenella longirostris*) and spotted (*Stenella attenuata*) dolphins often swim with yellowfin tuna in the tropical Pacific. Huge tuna seiners herd and set nets on the mammals to capture the fish beneath. In the process thousands of dolphin drown, though fishermen may make extraordinary efforts to free them. Between 1960, when tuna fishermen switched from hooks to purse seines, and 1975 an estimated three to five million dolphins were killed in the nets. Subsequent to new U.S. laws, public opposition, and fishermen's efforts, the dolphin kill has plummeted.

After years of argument and experiment on ways to reduce the incidental toll, in 1984 the U.S. Congress established a quota of no more than 20,500 dolphin deaths annually. Once that number is reached, the American fleet must stop fishing on porpoise, as the technique is known. Onboard observers from the National Marine Fisheries Service, and stiff fines for violations, give the law real teeth. Unfortunately, the same precautions and penalties are not in place for most of the dozen other nations that field tuna seiners, though Mexico is moving in that direction.

Another, less selective scourge is the drift-net fisheries in the North and South Pacific. A fleet of 1,500 vessels from several nations lays an estimated 20,000 nautical miles of gill nets out every night, and sweeps everything larger than the mesh from the sea's top 25 feet every morning. This includes thousands of sea birds, dolphins and seals, as well as the fish and squid that are the nets' intended prey. Some countries, like New Zealand, have banned drift-net fishing from their waters. But most such fisheries are in international waters and beyond national regulation.

SEALS AND SEA LIONS

Seals are the most diversified of the pinnipeds or flipper-footed sea mammals, which include walruses. They are divided into two main groups on the basis of their ears. There are those with external ears, the most common of which are sea lions, and those without external ears, sometimes called the "true" seals. Eared seals can rotate their hind flippers so that they walk on land with basically a four-footed gait. The others are obliged to inch along like a caterpillar or drag themselves by their front flippers.

Seals and sea lions are at home in the water, even though the land is where nature dictates they court, mate and give birth. Weaning is a rapid, often abrupt, process among the pinnipeds, ranging from two to twelve weeks. This places a premium on rapid blubber buildup. While detailed studies are not available on all seals, the Atlantic gray (*Phoca vitulina*) shows how crucial milk composition is to a mammal that must leave the land for the frigid sea in the shortest possible time.

Even after a year's gestation, the gray seal pup is born with a relatively thin layer of blubber on its thirty-pound body. Yet the mother weans it in approximately two weeks. To accomplish this, *Phoca*'s milk contains 53 percent fat at birth. By comparison, the California sea lion (*Zalophus californianus*) has 37 percent fat in its milk. Drinking the richest milk in mammaldom, the gray pup adds about four pounds a day to its birth weight.

When the pup is around 90 pounds, the mother gray calls it quits. She has lost comparable weight, dropping an average of 86 pounds in three weeks, and it is time to renew her own resources. After she disappears in the waves, the pup awaits her return in vain. Following a week's fast, the pup usually swims away from its beach-front nursery to catch meals as best it can.

By contrast, mother Weddell seals (*Leptonychotes weddelli*) are protective of their pups, which are twice as large at birth as the gray's, and nurse them six or seven weeks. These Antarctic seals learn to catch squid and octopus as well as fish. They are adept at using their teeth to keep open air holes, but the ice's movements kill some, as do orcas and leopard seals.

Some seals have had the additional burden of providing humanity with food and useful materials, like their skins or blubber. While fur seals were decimated for their pelts, no pinniped species has suffered more for being fat than the northern elephant seal (*Mirounga angustirostris*).

THE ELEPHANT SEAL

By 1870 the once-crowded island breeding grounds from Point Reyes north of San Francisco to the Los Coronados off San Diego were virtually bare of elephant seals. Hunting of what was estimated to be a few hundred survivors was unprofitable. Collectors went to Guadalupe Island, 180 miles from Baja California's arid coast, and shot the rare mammals for museum display cases.

Then the Mexican government placed the seals under partial protection in 1911. Eleven years later, even as revolution convulsed Mexico, the ban on slaying elephant seals was made complete and permanent. Within two decades elephant seals were again mating on San Miguel Island in the Santa Barbara Channel.

Researchers think that from Baja's original Guadalupe colony have come the elephant seals that now populate the isles off San Diego and Santa Barbara, and reach at least as far north as Ano Nuevo Island, near Santa Cruz, California. This is a very small source of genetic material for the more than 30,000 estimated living along the West Coast. Since 1972 they and other marine mammals have had federal protection in the United States.

Believed to be solitary and usually docile, northern elephant seals flock by the thousands to favorite rookeries in December to commence a ten- to twelve-week breeding season. All the adult bulls are spoiling for a fight. While they pick a spot on the beach to gather harems of around 30 females, beachmaster bulls defend their potential mates, not territory, against cruising bachelor bulls.

Like humans, this mammal has a prolonged gestation—roughly eight months—and usually gives birth to one offspring at a time. There is also a high pup mortality in the rookeries. Observers estimate that as many as half of the pups do not survive the first six weeks until they can be weaned. Most fatalities occur when pups are crushed by bulls rushing to defend their harems from challengers.

The females give birth in January and, because of a natural mechanism that delays implantation of the fertilized ovum, all within a short time of each other. Baby elephant seals are curly, black bundles of fur and flippers weighing from 35 to 70 pounds at birth. In four to six weeks of high-protein, high-fat nursing, the pups grow to 250 to 300 pounds. Then mother abandons them.

PENGUINS ARE NO STUFFED SHIRTS

All 17 of the world's penguin species live south of the Equator in tropical to frigid waters. The endangered Humboldt or Peruvian penguin (*Spheniscus humboldti*) resides in dwindling rookeries off Chile and Peru. The small, red-eyed rockhopper (*Eudyptes crestatus*) breeds by the millions on steep hillsides of sub-Antarctic islands like the Falklands, where Argentina and Britain fought a war in 1982.

By contrast, the 3-foot-tall king penguin (*Aptenodytes patagonica*) prefers places like South Georgia Island, which has shorelines and vegetation. The larger, heavier, and related emperor penguin (*A. forsteri*) breeds on sea ice along the rims of Antarctica. Though it may walk 100 miles over ice to get to its rookeries, the emperor never sets foot ashore. On the other hand, the hyperactive Adélie (*Pygoscelis adeliae*), the only other penguin actually living on the ice continent, seeks land in order to build nests of loose stones.

Neither emperors nor kings deign to build nests. They incubate their eggs by balancing them atop large, webbed feet and holding the eggs insides pouches formed by a fold of abdominal skin. But there the similarities end. Kings establish territories and both mates take turns during the southern summer or fall.

Emperors breed in the Antarctic winter when temperatures may fall to 80 degrees below zero, winds may reach 120 miles an hour, and it is dark 24 hours a day. After laying the egg, the female treks back to open ocean, which is usually miles away, to feed and regain her strength. The male is literally left holding the egg for three or four months. During this period he lives off his fat, frequently losing up to half his body weight.

When mother returns to take care of the newly hatched chick, father departs for what is probably one of the greatest food binges in the avian world. Diving nearly 900 feet and staying submerged for almost 20 minutes make the adult emperor a formidable predator, mainly of krill. Indeed, penguins and the crabeater seal (*Lobodon carcinophagus*) are considered to have greatly expanded their numbers since whalers reduced the natural competition for Antarctic krill. Now these animals face the pressure of human krill harvesters.

EXPLORING THE OPEN OCEAN

Outside the general coastal zone, which includes the polar continental shelves as well as those of the Atlantic, Pacific and Indian ocean basins, is the other 80 percent of the world ocean. For much of humanity's nautical history this Brobdingnagian expanse was feared and often considered accursed. Ships commanded by foolish captains fell off the ocean's edges. Sea monsters, presumably with a craving for sailors' flesh, populated the unknown waters, but little else lived out there. Even after superstition fled before explorers and navigators, it still seemed to educated people that the open ocean was a biological desert.

Then scientific study of the marine environment took hold and, even with relatively crude equipment by today's standards, started upending popular notions. It turned out that, except for large pockets in all three ocean basins, many different animals are distributed throughout the breadth and depth of the world ocean. The most numerous animals are the various zooplankton species, with copepods the single largest group.

Though they may be smaller than a grain of rice, copepods' segmented bodies, paired antennas and jointed legs recall their larger relations, the lobsters and shrimps. The 750 pelagic species of copepods range from the ocean surface down almost five miles, but the deeper the sampling the fewer the animals. If they are living on or near the surface, copepods of the tropical seas are a deep blue hue, which is also the color of these waters. Blending in with their environment affords these crustaceans some protection from sea birds and, perhaps, the strong ultraviolet radiation from the sun.

Not far below the surface other copepods are transparent or colorless. Deep sea species are most often tinged with areas of red; a few have black on their exoskeletons. This helps to camouflage them during vertical migrations from the 800- or 1,600-foot levels. Some dark-dwelling copepods produce light chemically. Many other animals living in the sunless midwater regions are also bioluminescent.

Among the surface drifters are the blue-tinged Portuguese man-of-war (*Physalia*) and sailor-by-the-wind (*Velella*) as well as the purple sea snail (*Janthina*). The first two are jelly animals that manufacture gas-filled internal floats, which catch the winds like the sails on a ship. Both jelly animals fish by trailing tentacle nets, but *Velella*'s shorter reach gives it less choice of prey.

Hanging on to the underside of the blue-tinted sailors, the tiny snail gradually nibbles away the fringe of this drifter. When *Janthina* is finished it floats away on a raft of bubbles. Shorn *Velella* bodies have been spotted anchoring juvenile gooseneck barnacles (*Lepas anatifera*), and if that morsel doesn't end up in some fish's stomach, the barnacles' weight sinks it into the abyssal ooze where bacteria finish the recycling. Nothing organic goes to waste in the oceanic scheme of things.

FORMIDABLE WIMPS: THE JELLYFISH

The lion's mane jellyfish (*Cyanea capillata*), or "hairy stinger," has a main body 20 inches or more in diameter. The size of its bell places it among the giant jellyfishes. Eight bunches of tentacles, each with around 150 individual strands, are spaced along the bell's rim. These pale yellow and orange-flecked strands extend 130 feet; when contracting they shrink to one-tenth their length within a second.

A cousin found only in the Arctic Ocean is believed to cover 500 square yards with its tentacles, making it the largest of the giants. Probably the most potent of the cold-water true jellyfishes, *Cyanea* frequents the northern regions of the Atlantic and Pacific oceans, but has been seen in the Baltic and North seas, too.

The exact nature of jellyfish toxin, which is generally not lethal to humans, is still a mystery. But the delivery mechanism is basically understood. Like their cousins the sea anemone and hydroid, jellyfish use layers of nematocysts in their skin to paralyze and kill prey or defend themselves against predators. These stinging capsules are not all alike. In jellyfish, they have become adapted to grabbing, holding and stinging the target.

When triggered by touch, chemical stimulation, or the vibrations of a swimmer, the oval-shaped nematocyst suddenly contracts, expelling various projectiles. One is a filament that wraps around any protuberance it hits. The thread may also be coated with a natural glue to adhere to the target or a corrosive substance, but it is the stinger-type of capsule that penetrates and injects the venom.

The stinger nematocyst functions much like modern whaling harpoons: A barbed tube hits the prey with enough force to rip open the armor of a small crustacean. Poison is released into the wound through pores in the tube. Outside tentacles contract, lifting the meal to the jellyfish's mouth, where special oral tentacles guide the food to the internal cavity for digestion.

The common jellyfish (*Aurellia aurita*) is found in most temperate and tropical waters and has such weak nematocysts that they cannot penetrate human skin. However, this disk-shaped, transparent jellyfish with no fringe of tentacles is potent enough for its own needs. An inch-wide animal, little more than a month old, can capture a fish 5 inches long. Within three months, *Aurellia*'s bell is 8 inches across.

The Portuguese Man-of-War is a prime example of a free-floating colony of individual organisms that perform different functions. These polyps are organized into a cooperative system: Some units keep the animal afloat and others allow it to capture, kill, feed and reproduce for the survival of all. Siphonophore is the generic name for 160 species of these marine animals, though they are sometimes collectively oversimplified with the tag of colonial jellyfish.

There are major differences between siphonophores and true jellyfish. Siphonophores usually drift, while jellyfish constantly swim. Most jellyfish species keep the sexes separate, releasing eggs and sperm through their oral cavity to unite in surrounding waters. Siphonophores reproduce by the budding of special units in the colony. The new main polyp breaks off and floats away.

The man-of-war's founding polyp can grow to 12 inches long and is crowned with a gas bag of equal length. An inflated, comblike structure atop the oval gas bag serves as a sail for this nomad of the world ocean. An unusual gland produces the gasses, which are mostly nitrogen, oxygen, and argon. Its tentacles range from 9 to 160 feet long and are armed with powerful nematocysts.

Surprisingly, Medusa fish and man-of-war fish swim within *Physalia*'s flotilla. The man-of-war fish not only scavenges uneaten food from its hydroid host, it also eats the animal's polyps. Juvenile jack fish and pompano swim near or among the tentacles of species like the Pacific purple-striped jellyfish (*Pelagia panopyra*). In the eastern Atlantic and North Sea, month-old codling not an inch long live among the strands of the lion's mane, and even help the jellyfish by browsing on parasites.

Some scientists suggest that such fish develop an immunity to coelenterate poisons; a few think that a protective mucus on the fish reduces its chances of being stung. Whatever the mechanism, it is not always reliable. On occasion, these brash hitchhikers feed the jellyfish with which they swim.

SHARKS AND OTHER FISH

Fishes are open-ocean marine life that come in various forms. For starters, there are the spindle-shaped flying fishes (*Exocoetidae*), which can lock open their winglike pectoral and pelvic fins and become airborne after breaking the surface at swimming speeds of 20 miles an hour. They have been reported gliding 25 feet at a time, pushed by rapidly vibrating tail fins.

On the other end of the shape spectrum is the giant sunfish (*Mola mola*), an oval-shaped flat fish with two extended fins, large eyes, but no real tail. Its body stretches nine feet and may weigh a ton. The record holder is a behemoth that collided with a boat off Sydney, Australia, in 1908, knocking out one of the engines. After limping into port, the boat and the sunfish were separated. The fish weighed in at 4,900 pounds and was nearly 14 feet long.

Sunfish are most often spotted basking or feeding on the surface. Favored prey are the many gelatinous animals, like salps, *Physalia*, and other jellyfish that are slow enough to be caught. Sun fish have been seen sucking and nibbling on jellyfish as they circle their prey. Both its size and pelagic ways make this saucer-like fish an ideal host for remoras (*Discocephali*), the suckerfish whose dorsal fin has been modified to form a vacuum chamber.

The swimming abilities of differently shaped sharks vary but include some of the ocean's fastest fish, none of which have swim bladders. Speed and maneuverability, crucial to catching rapid prey like mackerel and tuna, increase with buoyancy. Sharks augment buoyancy by reducing body density—their internal skeleton is of cartilage rather than bone—and developing large, oily livers. The liver of a blue shark (*Prionace glauca*) accounts for up to 20 percent of its body weight; some deep-sea sharks, like the dogfish clan, have proportionately larger livers.

Coastal hammerheads, which prefer warm waters and a squid diet, have gained extra lift from their flattened head structure. It acts like a wing in facilitating rapid vertical and banking movements, as well as providing superior binocular vision.

Slower, seabed-dwelling sharks tend to be small, with most warm- and cold-water animals growing no more than 6 or 7 feet long. (An exception are the tropical nurse sharks, which may be twice that length.) The active swimmers are routinely 10 or more feet long. In the case of the world's largest fish, the unaggressive whale shark (*Rhincodon typus*), individuals have been recorded at between 40 and 60 feet.

While all sharks are carnivorous, fortunately for swimmers and sailors the warm-water whale shark eats only small fish and plankton. Similarly, the second-largest shark in the sea, the 30- to 50-foot basking shark (*Cetorhinus maximus*) of cold temperate waters, is also a plankton grazer. The same cannot be said for the great white shark (*Carcharodon carcharias*), a basically coastal predator which grows to more than 20 feet in length and has been found far out at sea.

Just as the great white ranges widely through tropical and temperate seas, so, too, does its diet vary: seals, sea lions, dolphins and sea turtles, large bony fish, other sharks, sea birds, and carrion from dead whales. While humans have been attacked and occasionally eaten, they do not seem to be a regular part of *Carcharodon*'s diet, much less a preferred hors d'oeuvre. That is not to say great whites or other large sharks, like makos, reef or hammerheads, are welcome swimming partners. Some reef sharks, for example, are known to be territorial and may attack human intruders as well as other sharks.

Sharks are justly reputed to have extremely fine powers of smell—they can detect one part of blood in 100 million parts of water—and sensitivity to movement or vibrations in water. Even weak electrical currents enable them to locate prey or, for migrating species, possibly orient themselves to the earth's magnetic fields.

ANIMALS OF THE MIDWATER

Between the surface zone and the bottom is the midwater column. Since the average depth of the world ocean is over two miles, the midwater region comprises the greater portion of the sea's volume. Yet, of the ocean's principal zones, it is the least explored and its inhabitants' interactions the least understood.

For the greater part of a century, scientists considered this region featureless. Most animals retrieved were netted from surface ships in blind or sonar-aided sweeps; they were usually categorized as residents or transients, based on the frequency of their occurrence at the same depths. Few paid much attention to the gelatinous goo collected in fine-mesh nets along with diverse zooplankton; there wasn't much left to study. But in the last twenty years these views have changed dramatically.

Many of our concepts of what animals live in the midwater and how they do it are undergoing drastic revision as marine scientists study this region with increasing technical sophistication. Helping them probe the water column are small, manned submersibles, and the electronic eyes and arms of remotely operated vehicles (ROVs). The results have opened new lines of inquiry—and introduced a few new hazards to ocean exploration.

Woods Hole Oceanographic Institution's Robert Ballard, discoverer of the *Titanic* and deep-sea vent animals, once was attacked by a swordfish while descending in a submersible, the *Alvin*. Apparently attracted by the glowing view port of the sub, the large fish exploded out of the gloom at 1,700 feet and rammed the 25-foot-long craft. It's pointed beak penetrated a seam and broke off. The dive was canceled and the crew surfaced with two-and-a-half feet of bill sticking out of *Alvin*. Happily, damages were minimal.

Subsea observations by manned vehicles and ROVs have demonstrated that surface ship-based methods underestimated the numbers of creatures that inhabit this part of the realm, particularly the soft-bodied animals. "There are a lot of animals in the midwater column, more than we previously thought, and many of them have tentacles," says research biologist Bruce Robison. A principal investigator for the Monterey Bay Research Institute in northern California, Robison asserts, "The ecological equivalents of rocks and trees do exist in the midwater."

He and other researchers, such as Alice Alldrege of the University of California, Santa Barbara, and Harbor Branch Oceanographic Institution's Marshall Youngbluth and Richard Harbison, are documenting the importance of the once seemingly insignificant organic debris floating in the water column. This detritus, or marine snow, is found throughout the world ocean.

Marine snow consists mostly of dead plant material bound by mucus strands to which other organic particles stick during the long, slow drop to the bottom. It makes up a nutritious package. Beyond feeding bottom creatures, detritus forms inch-long habitats for bacteria and protozoans during the descent. In an ecological sense, these are edible rocks.

But the midwater ecology boasts larger, if harder-to-see, structures. Researchers are discovering that around 300 feet below the surface, where stability and calm prevail, life can flourish on even the most fragile of surfaces—like the mucus arrangements fabricated by tiny jellylike animals. From their vantage point in submersibles, scientists are seeing sheets, globes and strands of translucent jelly that range in size from less than an inch to around a foot. Many of these are discarded traps for filter-feeding animals less than half-an-inch long. They are a version of Robison's midwater trees.

Consider the larvaceans, a group of gelatinous tunicates that builds some of the mucus traps. These tiny, transparent plant-eaters strain food from seawater by catching particles in a mucus "house" the animal spins. The house, which varies in size and complexity from species to species, is much larger than the animal producing it, and is actually a filter chamber (or chambers). By lashing their tails the larvaceans in the *Oikopleura* family draw seawater through the filter chamber into a sievelike system that funnels particles into the animal's mouth.

During feeding, pressure rises inside the spherical house until a trap door springs open and releases a jet of water. Propelling themselves in bursts, the midwater *Oikopleura* move among the sea's floating pastures, collecting the smallest forms of phytoplankton known—ones less than a micron wide that easily elude manmade nets. In fact, studying larvacean houses was how scientists discovered the existence of these kinds of plankton.

Oikopleura have been taken from the open-ocean surface zone and the deep sea, but they and other gelatinous animals seem most abundant between depths of 800 to 3,250 feet. These jelly zooplankters tend to arrange themselves in narrow depth zones that reflect preferred feeding areas. This may mean there is "considerably more community structure in the pelagic environment than suspected," Marsh Youngbluth says.

When *Oikopleura* or other filter-builders discard their houses—some individuals can do this five to ten times a day—they usually leave bits of food stuck inside. Copepods and other tiny crustaceans often ride these drifting houses, which can persist for hours, and finish off the meal. Sometimes the passengers board before the house has been jettisoned. Salps, siphonophores, and other gelatinous animals also bear commensal hitchhikers, often for a longer stretch. Broken siphonophore tails may stay days in the water column.

One group of specialized midwater dwellers are the hatchetfishes (*Argyropelecus, Sternoptyx*), which are small, short, thin fishes with arrangements of reflective scales and luminescent cells, called photophores. Seldom more than 3 inches long, hatchetfish live below the 3,000-foot mark. They swim with their eyes permanently pointed upward to observe any prey silhouetted against the extremely dim light from above. Their reflective armor and internal lighting are concentrated along the bottom half so that their forms will be diffused to any predator below them.

Gulper and swallower fish live below 6,000 feet along with red squids (*Histioteuthis*), which have membranous webs between their tentacles, and a blind octopus (*Cirrothauma*). Gulpers, or pelican eels, range from 2 feet to 2 yards long. Most of this is the whiplike tail and the rest is a stomach with teeth. Homing in on their prey's lights, the gulper's jaws unhinge to take a bigger bite. Its belly expands to several times its usual size, allowing it to swallow a meal larger than itself.

Deep-sea anglerfishes are usually 2 to 4 inches long, though females of more than a foot in length are found.

These fish hunt by dangling a luminescent lure on the tip of a fleshy stalk in front of their open mouths. Anything unwary enough to swim up for a closer look is engulfed.

The problem of finding a mate has been solved by these residents of the midnight realm in an efficient, if parasitic, manner: Mature males swim freely until they encounter the much larger females. The male sinks his teeth into her side and holds on. Eventually, he degenerates so that once his circulatory system blends into hers he is reduced to a small sac holding only the male sex glands.

HITTING BOTTOM

To humans, the obvious features of the abyssal seabed are near-freezing temperatures, inky darkness and almost unimaginable pressure. The *Trieste*'s 1960 drop of nearly seven miles into the Challenger Deep placed a weight of more than eight tons on each square inch of the bathyscape. Yet a variety of fish, soft-bodied animals, and crustaceans have adapted to these conditions by equalizing external and internal pressure, allowing them to live in equilibrium.

Since food and oxygen are much less abundant than in the surface zone, the biology of bottom animals tends to run in slow motion. Fish trapped and monitored on the seabed consumed oxygen at a rate half that of shallow-water fishes of similar size. One aspect of a low metabolic rate is that food energy lasts longer—an advantage when meals are few and far between. This tends to keep animals small and long-lived, such as a one-third-inch clam taken from 12,350 feet that was calculated to be around a century old. Still, sleeper sharks (*Somniosus microcephalus*) ranging between 16 and 26 feet in length have been found nearly 5,000 feet down, and giant squid (*Architeuthis*), the sperm whale's preferred prey, are captured at such depths.

Largely unknown until two decades ago, many of the species collected from the sea floor are mysteries to science; some are still scientifically undescribed and await classification. Among the discoveries that have excited popular as well as scientific interest are the approximately 100 new species associated with hotwater vents along cracks in the ocean floor. These tube worms, clams, mussels, crabs, anemones and other animals are not the run-of-the-abyss resident. They have adapted to waters that are warmer than the normal 35-degree temperatures and suffused with concentrations of minerals, such as sulfur, hydrogen, ammonia, methane, iron, and manganese.

HYDROTHERMAL VENT ANIMALS

' In 1977, the research submersible *Alvin*, in the words of marine biologist James Childress, unexpectedly came upon "an oasis of densely packed animal life" 8,500 feet below the Pacific Ocean's surface, about 270 miles northeast of the Galapagos Islands. The oasis formed around a system of hot springs along a ridge where Earth's crust was spreading apart. Some of the vents took the form of chimneys pumping out dark plumes of minerals, which were labeled "black smokers." Robert Ballard and the other geologists on board had come upon one of those rare discoveries that reshape scientific thinking.

"Here were species previously unknown to science,

living in total darkness in densities enormously higher than had ever been thought possible in the deep sea," said Childress, who is with the University of California, Santa Barbara. He helped document the finds in a later dive. Giant red-tipped tube worms (*Riftia pachyptila*) a yard long, white clams (*Calyptogena magnifica*) a foot long, and yellow mussels (*Bathymodiolus thermophilus*) clustered around vents spitting out dissolved minerals, including normally toxic hydrogen sulfide. Found in smaller but significant numbers were shrimp, white crabs, and fishes.

Marine scientists were amazed at the size of many of the animals. The worms were particularly impressive because they built long white, one-inch diameter tubes of horny material which, in some cases, extended for yards. This may allow them to follow shifts in the flow of hydrogen sulfide from the vents while holding their filter-feeding tips constantly among the dissolved minerals. When clumped together, *Riftia*'s prominent red tips atop curved stalks gave observers the impression of flowers. A particularly dense patch along the Galapagos rift was tagged "The Rose Garden." But how could a diverse group of animals be so abundant more than a mile below the life-giving zone of light?

The answer turns out to be living better through chemistry. Unlike other benthic creatures, vent animals survive in an ecosystem based on sulfur-eating bacteria that thrive in the normally toxic, hydrogen sulfide-rich hot water. Strands and mats of bacteria were free-living around the vents. Most importantly, symbiotic bacteria were found in the tissues of tube worms and in the gills of clams and mussels. The microbes were breaking down the sulfide to sustain the animals' energy needs. This process is chemosynthesis instead of photosynthesis, though nutrients also come from organic particles in the sediment and water.

The vent animals adapted to living on chemical energy so efficiently that they are abundant in the midst of an apparent scarcity of other deep-sea inhabitants. Childress has observed that the hot springs "is one habitat in the deep sea where the density of life equals, if it does not surpass, what is found in any other marine ecosystem."

One of Childress's colleagues grappling with the complexities of the chemosynthesis puzzle is zoologist Charles Fisher. He notes the intriguing ways vent animals handle deadly chemicals like hydrogen sulfide. The giant tube worm and the rift clam bind the toxic sulfide to unusual blood molecules and transport it to the bacteria. Other animals, like crabs and many of the worms, except *Riftia*, coat their gills or whole bodies with substances that prevent the sulfide from entering their systems.

There are more questions than answers here, for not all the rift animals feed the same way. The yellow mussels, which also carry bacterial symbionts, apparently change the sulfide into a nontoxic form and transport it to their internal bacteria for digestion. These mussels also filter-feed on bacteria in the water.

Since the discovery of the Galapagos rift animals, similar communities have been found all over the world, including sites off Japan, south of the Baja California peninsula, and in the Atlantic. One was reported in 1988 only 125 miles off the Oregon coast at a depth of about 9,000 feet. Biologists are confident that as the search expands to other sea floor spreading zones more vent ecosystems will be identified.

Cold-water oil seeps in the Gulf of Mexico and off the Japanese coast suggest that chemosynthesis is not limited to hydrothermal vent communities. Clams and tube worms gathered off Louisiana from seeps 1,300 and 2,000 feet down held sulfur-eating bacteria like those in the vent animals. But an unclassified specie of mussel (*Mytilidae*) harbored methane-eating microbes in its gills. Their bacteria ignored sulfide and went for the methane in this hydrocarbon-rich environment. Tube worms living off sulfide have been found next to methane-eating mussels.

And yet the vents are the most dramatic and intriguing. In mid-Atlantic rift zones, black smokers are frequently the home of gray shrimp (*Rimicaris*) that have neither eyes nor the eyestalks usually found on their kind. These shrimp blanket the seabed around vents spewing forth 660-degree water, easily hot enough ·to boil the two-inch-long crustaceans. How do the shrimp, which eat the sulfide deposits, avoid the danger just inches away? They have the equivalent of eyes on their backs, Woods Hole biologist Cindy Lee Van Dover has discovered. These patches on their body armor are sensitive to the extremely low light generated by heat from the vents. Human eyes cannot detect light but shrimp can.

ISLANDERS
ON THE WORLD OCEAN

From seabed to surf zone the world ocean is under increasing pressure from an expanding human population. Today, we employ the ocean as a vast transport system, linking the continental islands that constitute only 29 percent of the planet's exposed skin. We suck petroleum and gas from offshore shelves, extract most of our salt, magnesium, and bromine from seawater, collect sand to build concrete landscapes, and range the ocean to gather fish, pelts, and, in some cases, whale meat and oils. Most commonly, we dump our cities' sewage and toxic wastes into it with little or no treatment.

An estimated 75 percent of America's 250 million people live within 50 miles of a shoreline. Industrial wastes combine with agricultural runoff and urban pollution to attack the plants and animals trying to survive in the coastal zone. Too often we forget that life on the land is directly linked to life in the ocean. Sometimes the sea reminds us by tossing back our wastes on to the beaches where we seek relaxation. In other instances the warnings are more subtle—and far-reaching.

Nautilus researcher Peter D. Ward provides a vivid illustration in describing changes in the Tanon Strait in the Philippines. Due to natural barriers on each end of the 145-mile-long strait, violent tidal currents are reduced, providing fish, corals, crustaceans, and fishermen with nearly ideal living conditions. The human population grew large upon the bountiful and beautiful strait.

The strait was a perfect habitat for the larger species of nautilus. The area had been heavily fished for the shelled mollusk at least since 1971, when a scientist had some 3,000 animals caught that year. When the shell of the chambered nautilus became popular among collectors, the strait's fishermen trapped an estimated 5,000 animals annually until shortly before Ward's 1987 visit. A week of intensive fishing yielded only three for Ward. The nautilus's disappearance is part of a larger story.

The islanders' preferred food fish, like the tuna, had disappeared in the 1960s and 1970s, along with the coral reef's giant clam. Sea turtles dwindled, and the salt water crocodile was decimated for its skin and meat. On land, thick rain forests were toppled and lowland forests replaced by rice fields. Pesticides washed into the strait, cutting marine productivity and bird life with the same chemical scythe.

To counter declining fish catches, fishermen turned to using dynamite on the miles of coral reefs. The Nobel Company helpfully moved plants into the region and the government encouraged this destructive technique. About 20 years later, nearly all the reefs lining Tanon Strait had been shattered, and with them went the nurseries of many fish species.

There are other, equally dramatic examples of unenlightened human activity in every coastal region of the world ocean. Major oil spills in the life-sustaining waters of Alaska and Antarctica within weeks of each other in early 1989 rang loud alarms. The eastern seaboard of America looks as biologically blighted from space as does the Mediterranean and the North seas, according to astronauts. In the largest die-off of its kind, more than 1,000 East Coast bottlenosed dolphins were killed in 1987 and 1988. Official reports blamed a natural toxin that poisoned fish the dolphins ate. However, the origin of that toxin may have been man-made pollution, other researchers claimed.

These are symptoms of major disruptions in the larger, interconnected fabric of life on Earth. Far more threatening to both humanity and planetary ecosystems are the signs of deterioration in the ocean-atmosphere relationship.

Recall that photosynthesis in the plankton-laden surface zone supplies a major part of the world's oxygen and uses up carbon dioxide as well. While seasonal holes in Earth's protective ozone layer have been confirmed above both poles, the one hovering over Antarctica has also been shown to diminish plant productivity on the sea surface. That reduces the number of zooplankton that can be supported and vibrates up a series of food webs. It also cuts back the amount of oxygen released into, and carbon dioxide taken out of, the atmosphere, factors in the apparent trend toward global warming.

Seawater protects plankton somewhat, so the penetration of ultraviolet radiation is much reduced below the ocean's top three feet. Unfortunately, the ultraviolet radiation is the strongest during spring in the Antarctic. This is the height of the most productive part of plankton's life cycle. While there is international agreement on banning by century's end the chemicals destroying Earth's ozone layer, these chemicals remain active for years after release.

We are all islanders on the world ocean. If we do not stop undermining our marine support systems, the day will come when the damage will have gone too far to be reversed. Even partial knowledge points to awful and tragic consequences if we try to ignore how connected we are to a truly wondrous living planet.

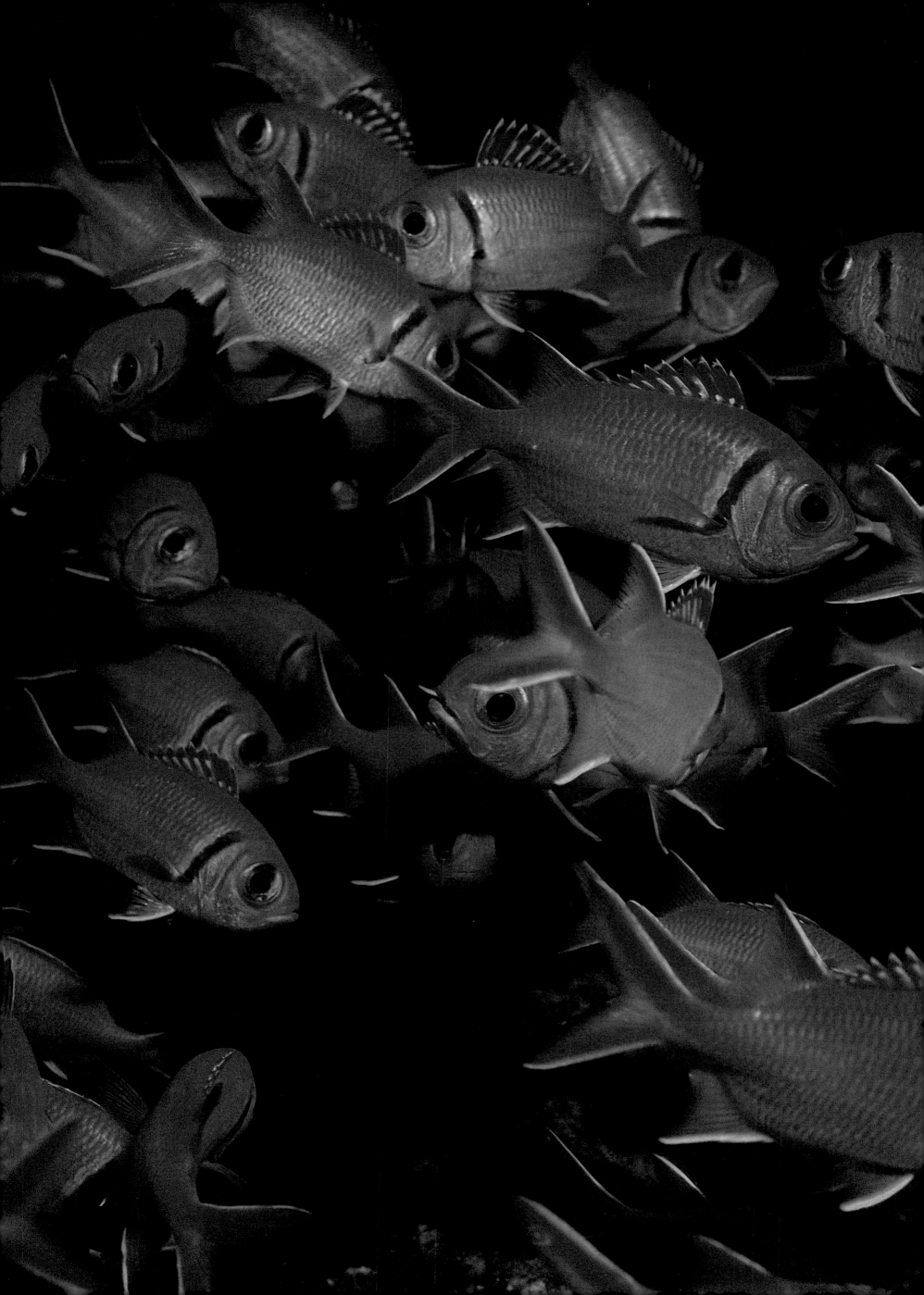

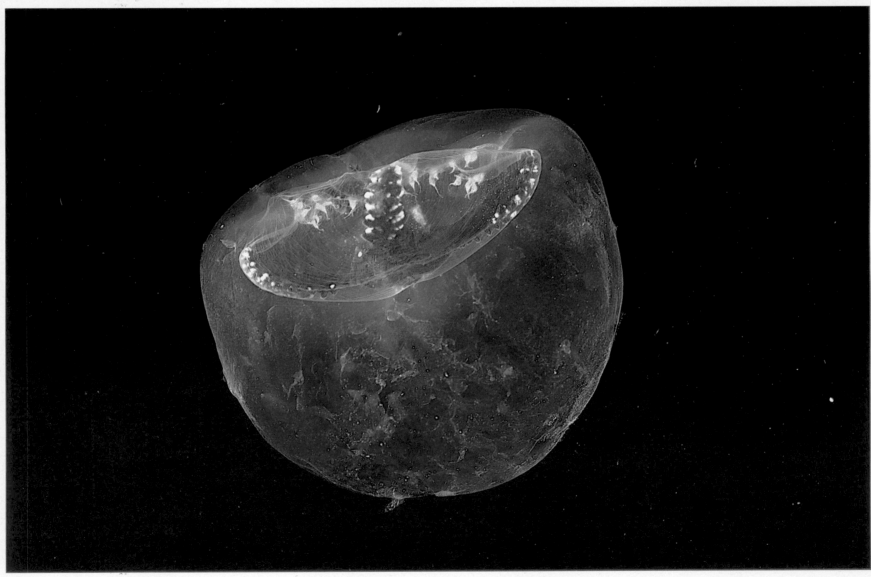

Not all jellyfish look the same: Polyorchis, a shallow-water, 3-inch-high resident of Pacific kelp beds (top), has red-colored light receptors, but a smaller, unnamed deep-water Medusa with dark yellow gonads (bottom) does not. The Pacific fang tooth fish (opposite), which lives in the dark below 2,000 feet, needs large eyes. Micronesia's beautiful 70 islands (overleaf) rest on limestone formations built by animals called coral polyps.

*A wolf eel (**left**) coils about itself; a midwater deep-sea anglerfish
(**top**) uses barbels on its head to lure prey within gulping range; the
moon jellyfish swims with rhythmic pulses of its bell. A classical
coral reef lagoon (**overleaf**) is Kayangel Atoll in Palua, Micronesia.*

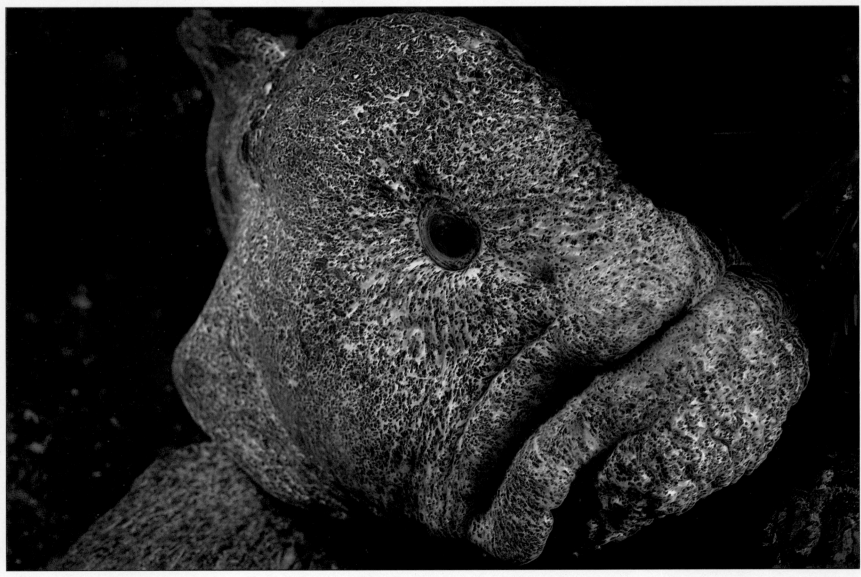

*Offering versions of fishy stares are an Australian scorpionfish (top),
a northwest Pacific wolf eel (bottom), a red Irish lord (right top),
and a fringe-headed blenny (right bottom). A banded pink shrimp
crawls on a pink Tealia anemone (overleaf).*

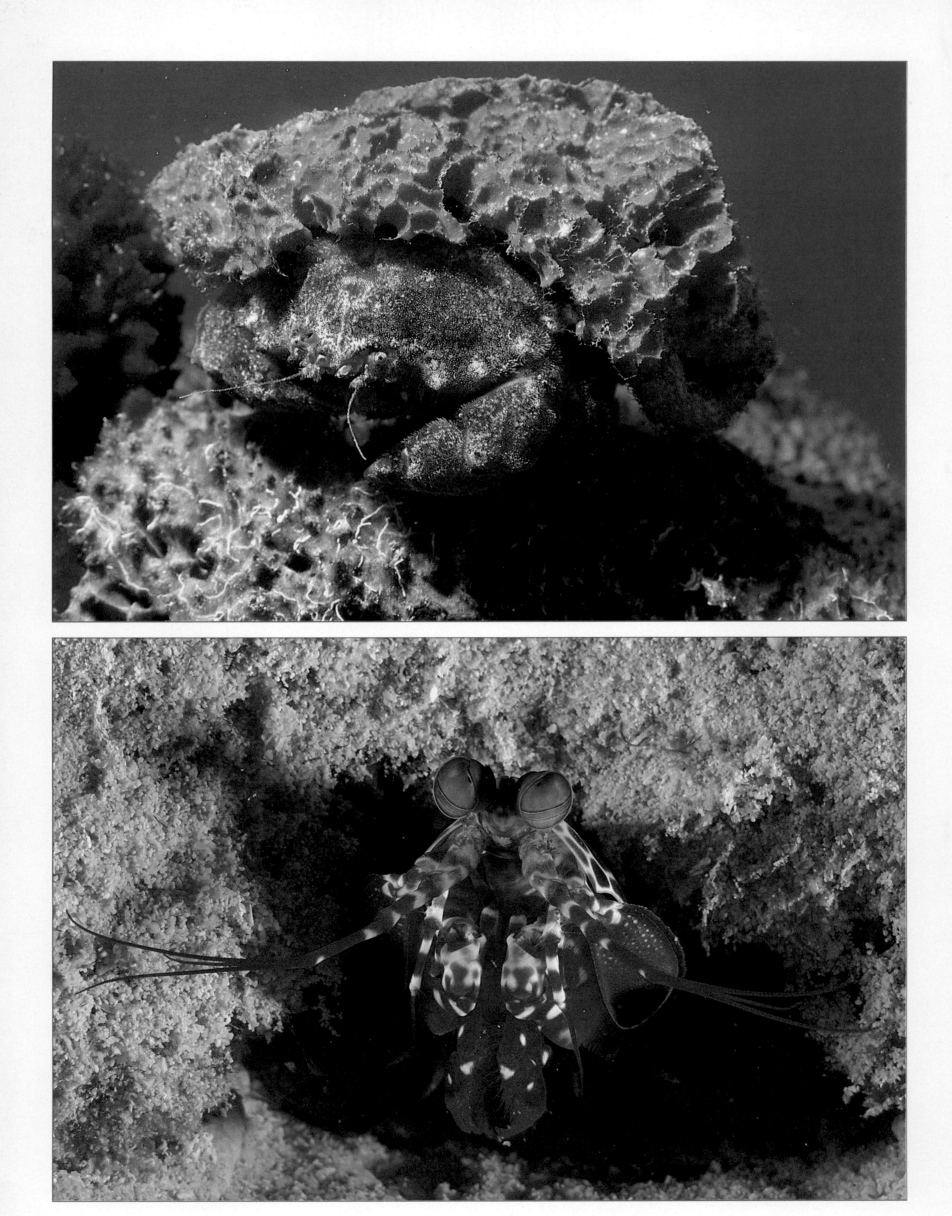

The Caribbean sponge crab (top) camouflages itself with a piece of sponge its back legs hold in place; a mantis shrimp (bottom) backs into its burrow; an orange ball anemone from the Cayman Islands (opposite) is loaded with stingers but very innocent-looking. The long-nosed hawkfish is on soft corals in the Red Sea (overleaf).

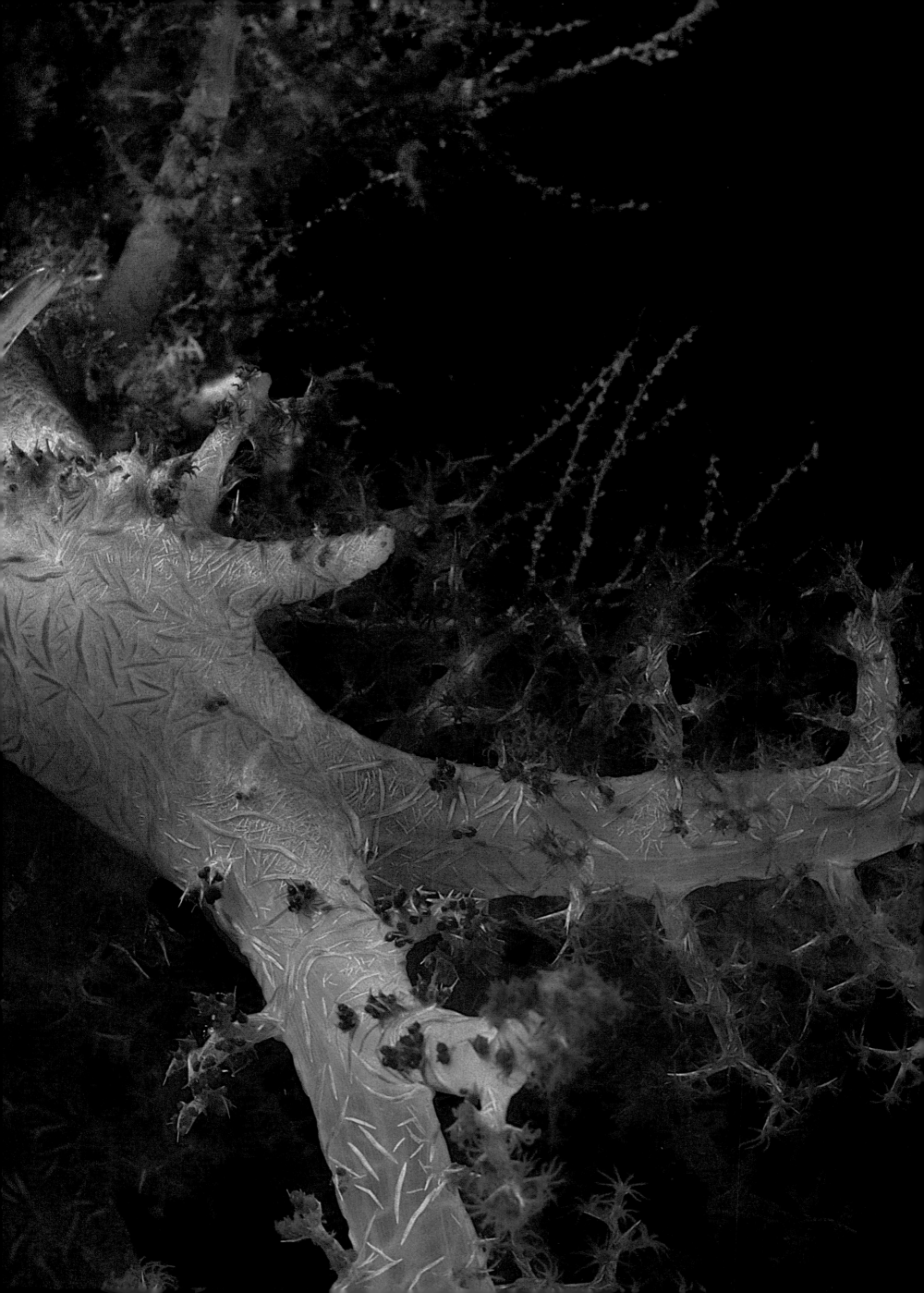

*Leafy sea dragons (**opposite top**) are Australian cousins of the common sea horse; a blue lobster (**opposite bottom**) is really a stage the American lobster goes through. The hawksbill turtle (**top**) depends on sandy beaches for its eggs, while the blue-spotted fantail stingray (**bottom**) uses sand to disguise itself. Barracudas school in the Red Sea (**overleaf**).*

A Hawaiian hawkfish finds comfort among yellow coral (opposite top), while a tropical sea urchin seems protective to some black fish (opposite bottom). The leather seastar crawls past an anemone (top), and a blood starfish climbs over purple hydrocoral (bottom). Enoplosus armatus, sometimes called "old wives" fish, are an unusual branch *of the coral reef angelfishes (overleaf).*

This spotted puffer (top) eats urchins, but holds still for cleaning by a wrasse; the longnosed butterflyfish (bottom) swims past orange cup corals. Stinging catfish (opposite top) gather along Japan's coast, and a spotted moray eel swims through soft coral (bottom). An Indo-Pacific jellyfish pulses upward in a spray of beaded tentacles (overleaf).

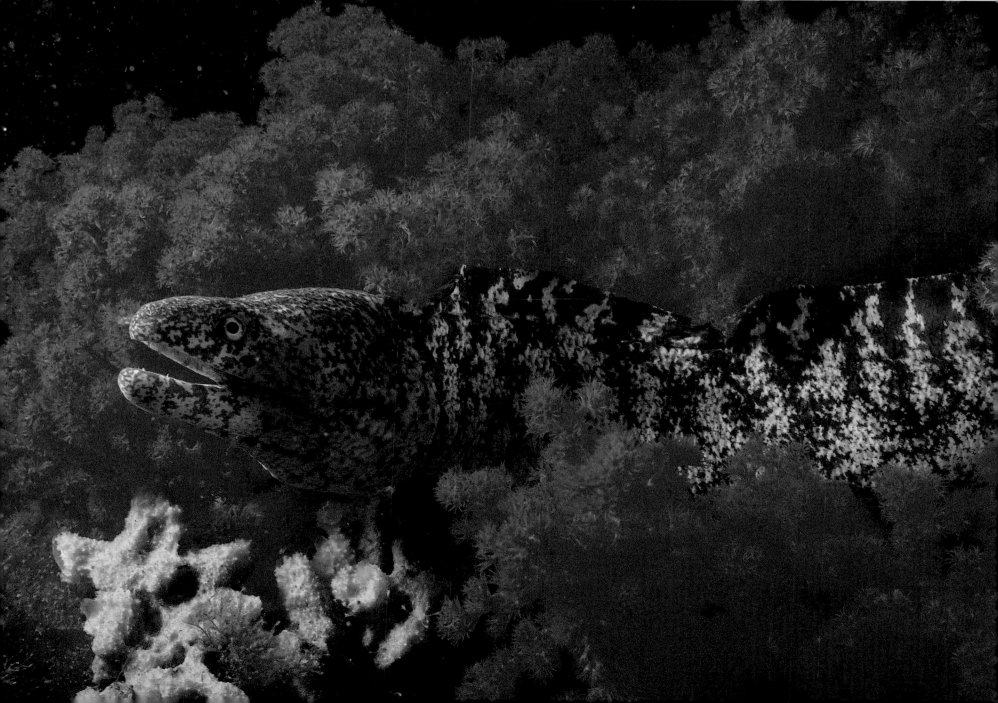

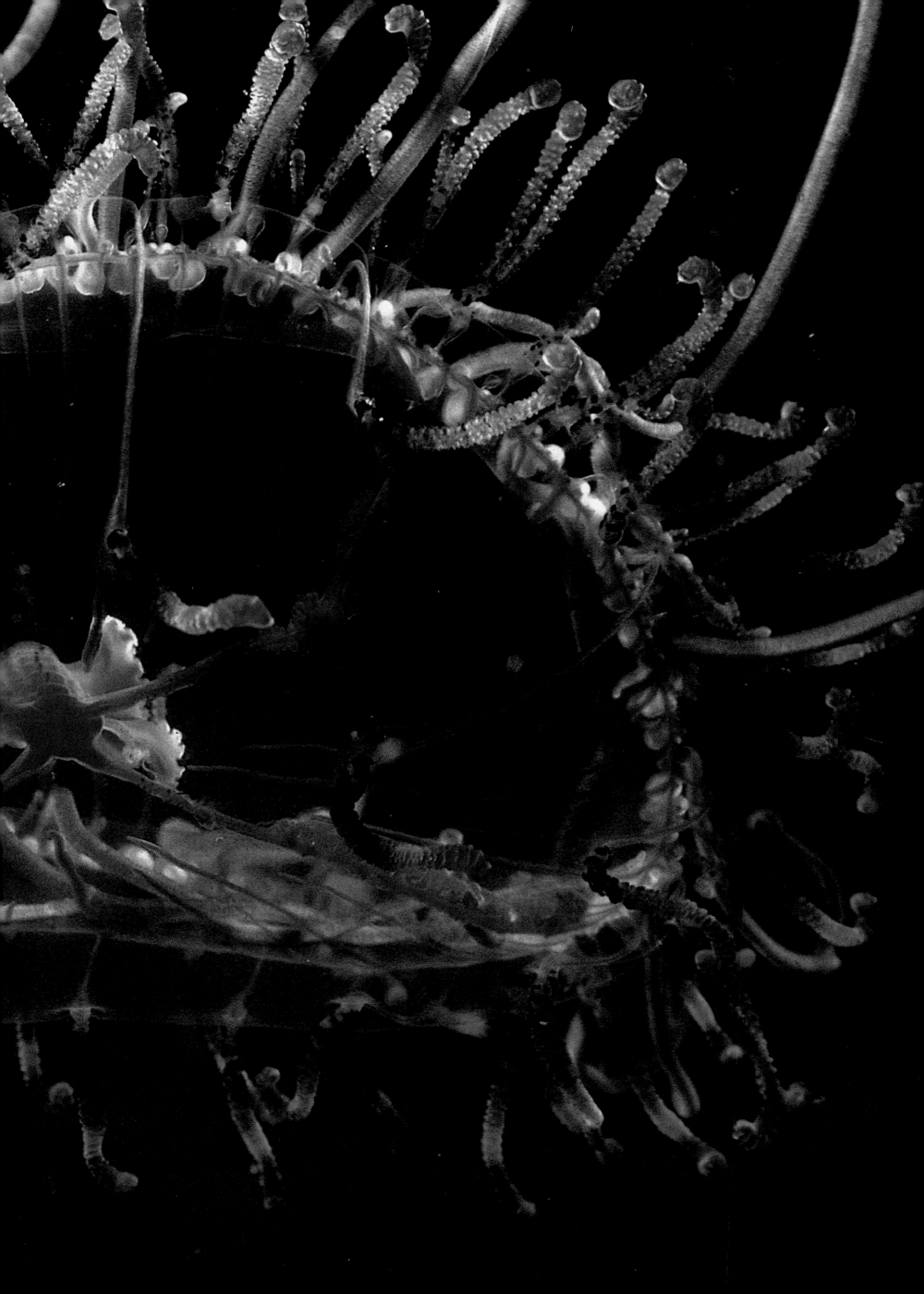

Anthias fish hunt on a coral reef (top); flame scallops sift the water with a tentacle net as they rest on a sponge (bottom). A giant green anemone, which may grow a foot wide, eats a small fish (opposite top). Water-powered scallops flee backwards from the sunflower starfish (opposite bottom). A great white shark emerges from the gloom (overleaf).

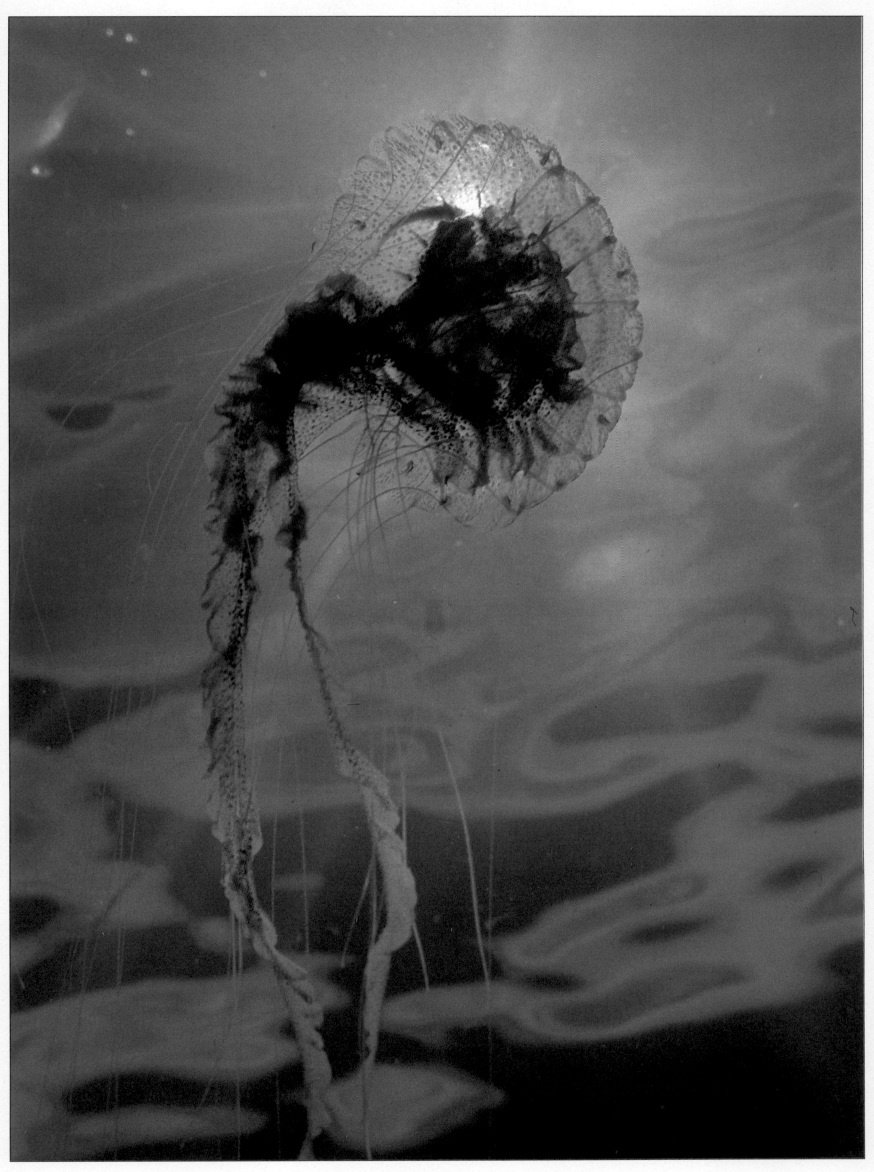

A small fish inside the jellyfish bell (above) may be treading the tightrope between being protected and becoming a meal, or it may have already fallen off. Bluebell tunicates on gorgonia take in sea water through the ringed siphon and expel it from the other (opposite top); the 2-inch-tall funnel tube worm extends its red gills (opposite bottom). Scalloped hammerhead sharks pass in an unusual display above a Sea of Cortez sea mount (overleaf).

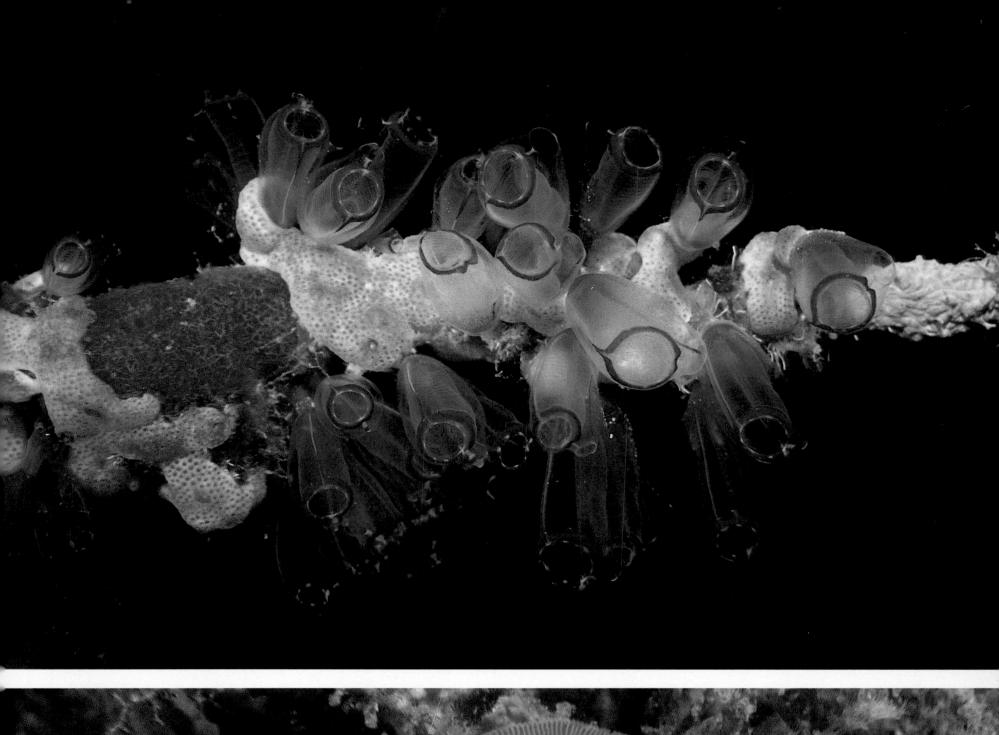

*Rays, skates and sharks are related, nonbony fishes. While the great white shark (top)
grabs food with its teeth, the Pacific manta ray (bottom) funnels plankton into its
mouth with its front fins. Basking sharks (opposite top) are 50-foot-long plankton-
eaters; blue sharks (bottom) are swift chasers. A pair of striped pilot fish accompany
a common oceanic whitetip shark (overleaf).*

54

An Indo-Pacific crown-of-thorns starfish, which eats coral polyps, regards a red gorgonia coral (top); a lemon nudibranch, or sea slug, breathes through the lacy tissues on its back (bottom). In the Galapagos Islands, a striped sea slug glides over orange tunicates (opposite top). A Nassau grouper stirs a flurry of herring (bottom) near Grand Cayman Island in the Caribbean.

The Galapagos Islands sea urchin's tiny pincers are held in white tubes among rows of spines (top). Azure vase sponges (bottom left) filter food by passing water through the sides; the spiral gill worm (bottom right) uses featherlike gills. A Spanish shawl sea slug's head resembles that of its garden snail cousin (opposite top). Giant red sea urchins of the Pacific grow to be 7 inches wide. Tropical tree corals (overleaf), a soft coral, are given shape by internal water pressure and hairlike reinforcements.

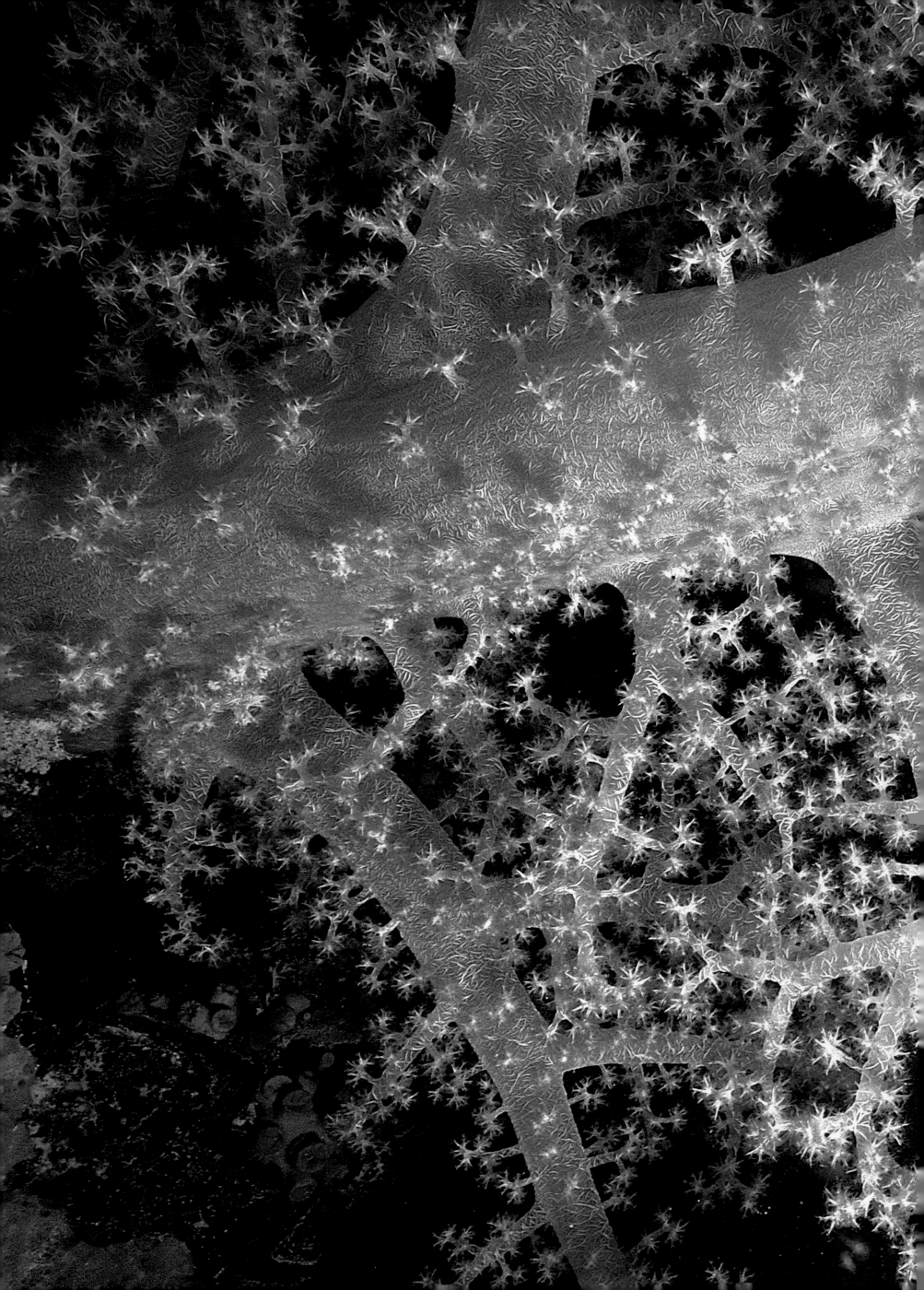

Copper, or whitebelly, rockfish (opposite top) are commercially important to Pacific fisheries. A parasitic isopod bites into and hangs on to a swimming soldierfish (opposite bottom). More helpfully, a tiny goby cleans a giant hawkfish's head (top). The stonefish (bottom) injects poison through its back spines. Blue sharks can grow to more than 18 feet in length (overleaf).

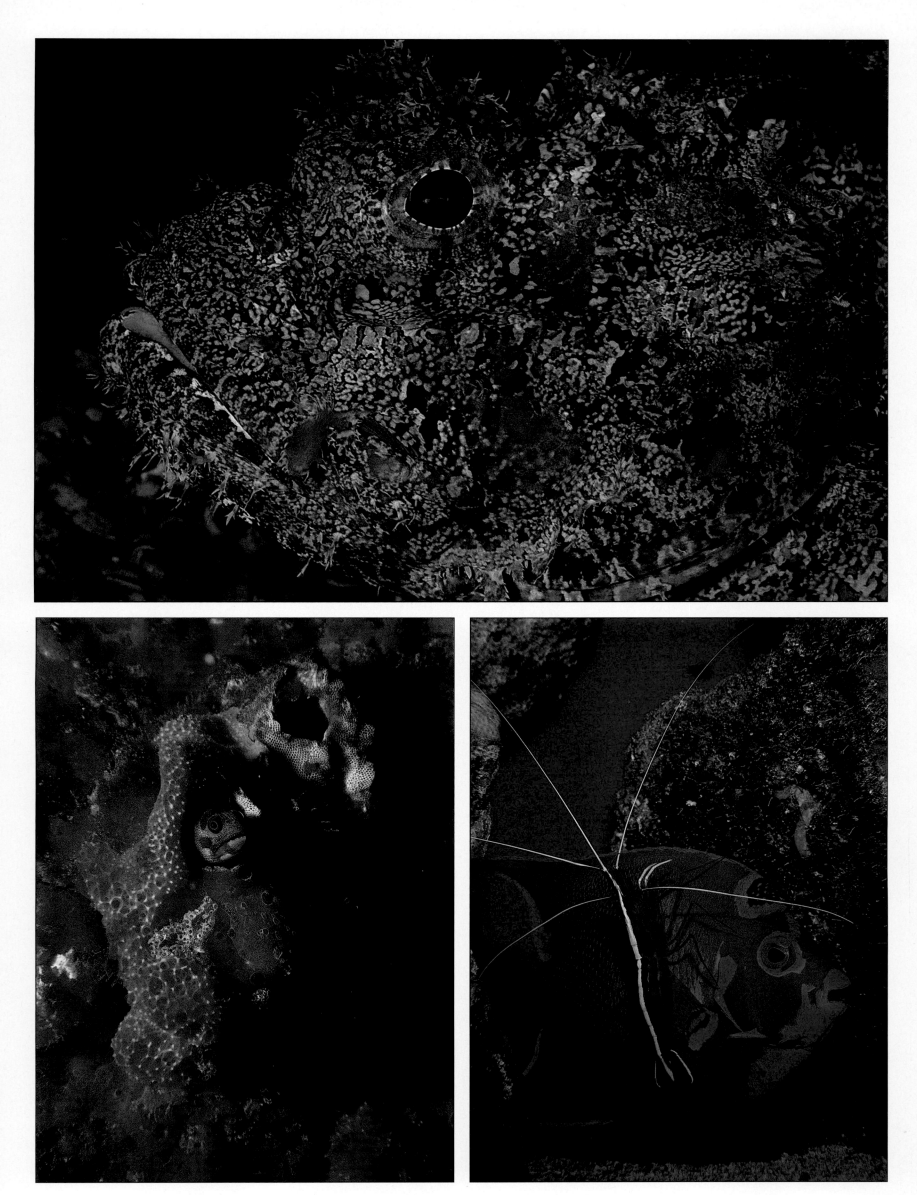

Stonefish (top) blend into lagoon debris, for which they are often mistaken. A blenny peeks out of sponge-encrusted barnacle remains (bottom left); a scarlet lady cleaner shrimp serves an angelfish on a reef (bottom right), while a grouper opens up for an inside cleaning job (opposite bottom). The stoplight parrotfish protects itself with a mucus cocoon when it sleeps (opposite top). Mako sharks are potential man-eaters, but far more sharks have been eaten by people than vice-versa (overleaf).

A slipper lobster walks over rocks in the Sea of Cortez (top); a horn shark (bottom), one of the older forms of shark, has stout horns in front of each dorsal fin; orange cup corals (opposite top) are shown closed and with feeding tentacles out. This hermit crab has occupied a murex snail shell (opposite bottom). A diver peers through a passing school of squirrel fish (overleaf).

A small pod of Atlantic spotted dolphins swim together off the Bahamas. The smaller one breaking formation is a juvenile; it develops more spots as it matures. These dolphins love to ride vessels' bow waves and are usually curious enough to get close to swimmers. Pacific spotteds are often trapped and killed in net fishing for yellowfin tuna.

Pacific Crevalle jack fish **(top)** *are important food fish; a large black grouper nears a
sea fan-lined reef* **(bottom)** *off Cozumel, Mexico. Remoras, or suckerfish* **(opposite top),** *attach
to sharks, rays, whales or even ships with a modified fin atop their heads that acts like
a suction cup. Rainbow runners school in the Sea of Cortez* **(opposite bottom).**

California spiny lobsters (opposite top) *make good eating, but the surgeonfish* (opposite bottom) *must be handled with care. It has sharp spines near the tail on each side. The tropical loggerhead turtle* (top) *rarely exceeds a yard's length, but that looks large to the cleaner coral shrimp* (bottom).

A soldierfish patrols its territory against a backdrop of feeding coral polyps (top); hoping it can pass as a finger sponge (bottom left), a slender trumpetfish stands at attention. The orange cup coral eats (bottom right), and a baby basket starfish wraps white-beaded arms around a filter-feeding gorgonia (opposite).

Bennett's nudibranchs mate while an indifferent brown worm inches past
(top); a feather duster worm (bottom) sweeps for food near corals; a flame
scallop, or file clam, feeds with tentacles (opposite top). When alarmed
the balloonfish inflates with water (opposite bottom), stiffening its spines.

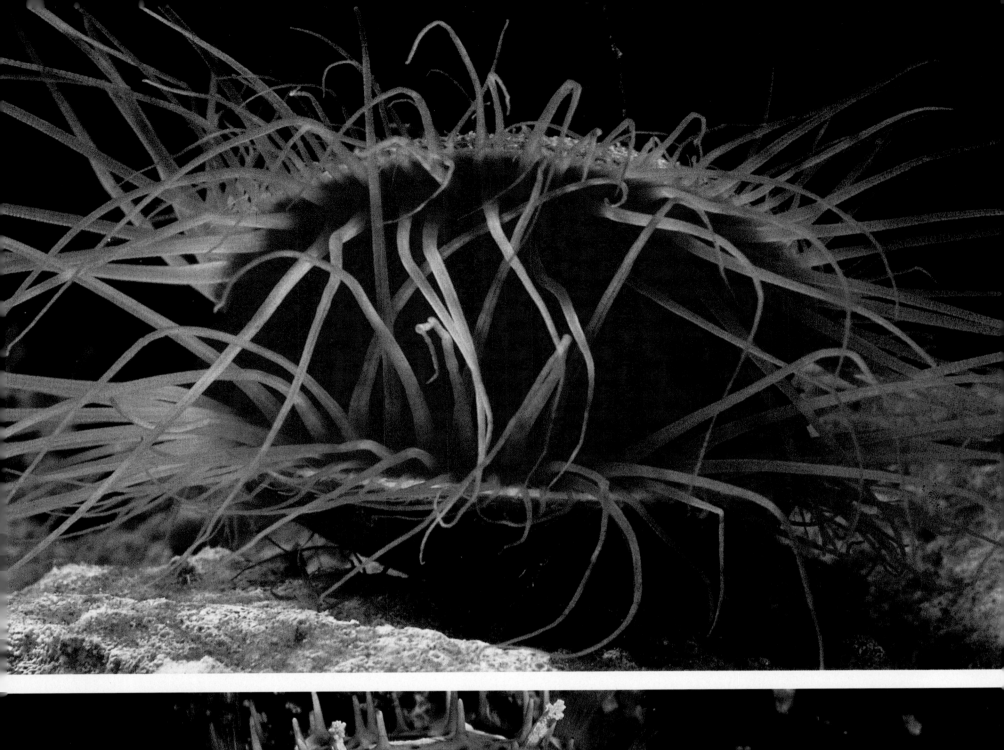

Maldive Island reefs harbor many colorful fish, including striped grunts (opposite).
Territorial disputes provoke threat displays (top), and may end in these grunts
fighting mouth-to-mouth. The thick spine on the lower gill edge of the queen angelfish
(bottom) is typical of salt water angelfishes.

*Clownfishes, or damselfishes, live among lethal anemone tentacles (opposite) for most
of their lives without being attacked. The cleaner wrasse removes parasites from
the blueface angelfish (top), which can crush shellfish in its jaws. Tunicates filter feed
(bottom) by drawing in water at the top and expelling it from the side; orange cup
corals just wave tentacles (overleaf).*

Moray eels suck water through their mouths and over their gills to breathe, making them appear either menacing, like the jewel moray (opposite), or comical, like old snaggletooth (above). Plumed lionfish melt into the seascape of encrusting animals on a wreck in Truk Lagoon (top); reef fish often rely on vivid bands of dark and light to evade predators (bottom right).

Stabs from stingray (top) tail spines are aggravated by venom, which has been blamed for some deaths. The great white shark is both a coastal and open-ocean predator (bottom), preferring fish and seals to humans. Schools of barracuda (opposite) will herd prey fish into dense shoals before they attack.

Not all anemones feed alike. A Caribbean carpet anemone (top) has so many ball-tipped tentacles the mouth is hidden, while the stalked anemones **Metridium** *(bottom left)* and **Tealia** *(bottom right) show the respective armaments of a delicate filter-feeder and a fish trapper. Soft corals (opposite) extend hairlike cilia to gather particles. Gorgonia coral is sheathed in horny material similar to fingernails (overleaf), and feeds in the dark. This one is growing on a giant clam in Truk Lagoon in the Carolinas.*

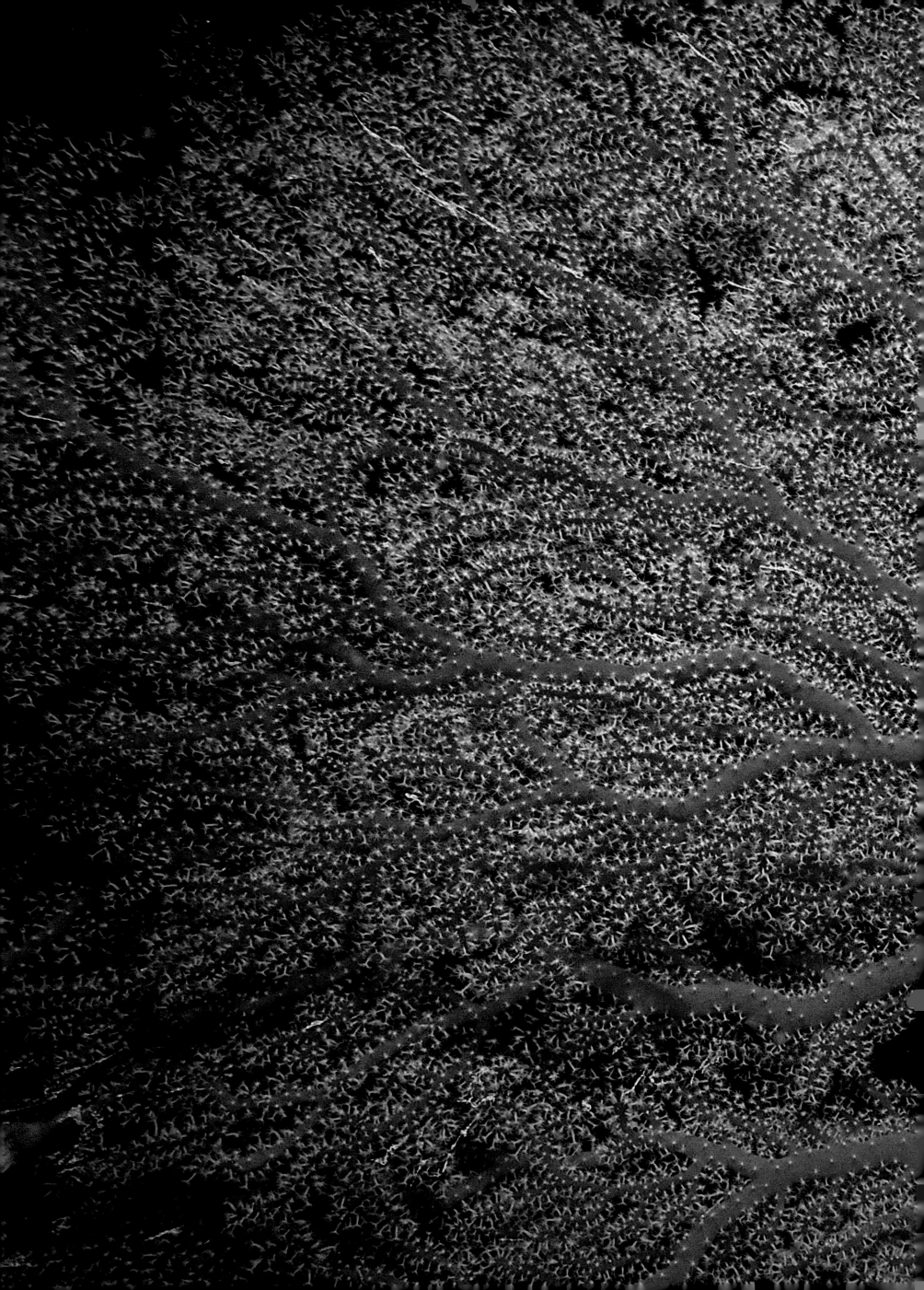

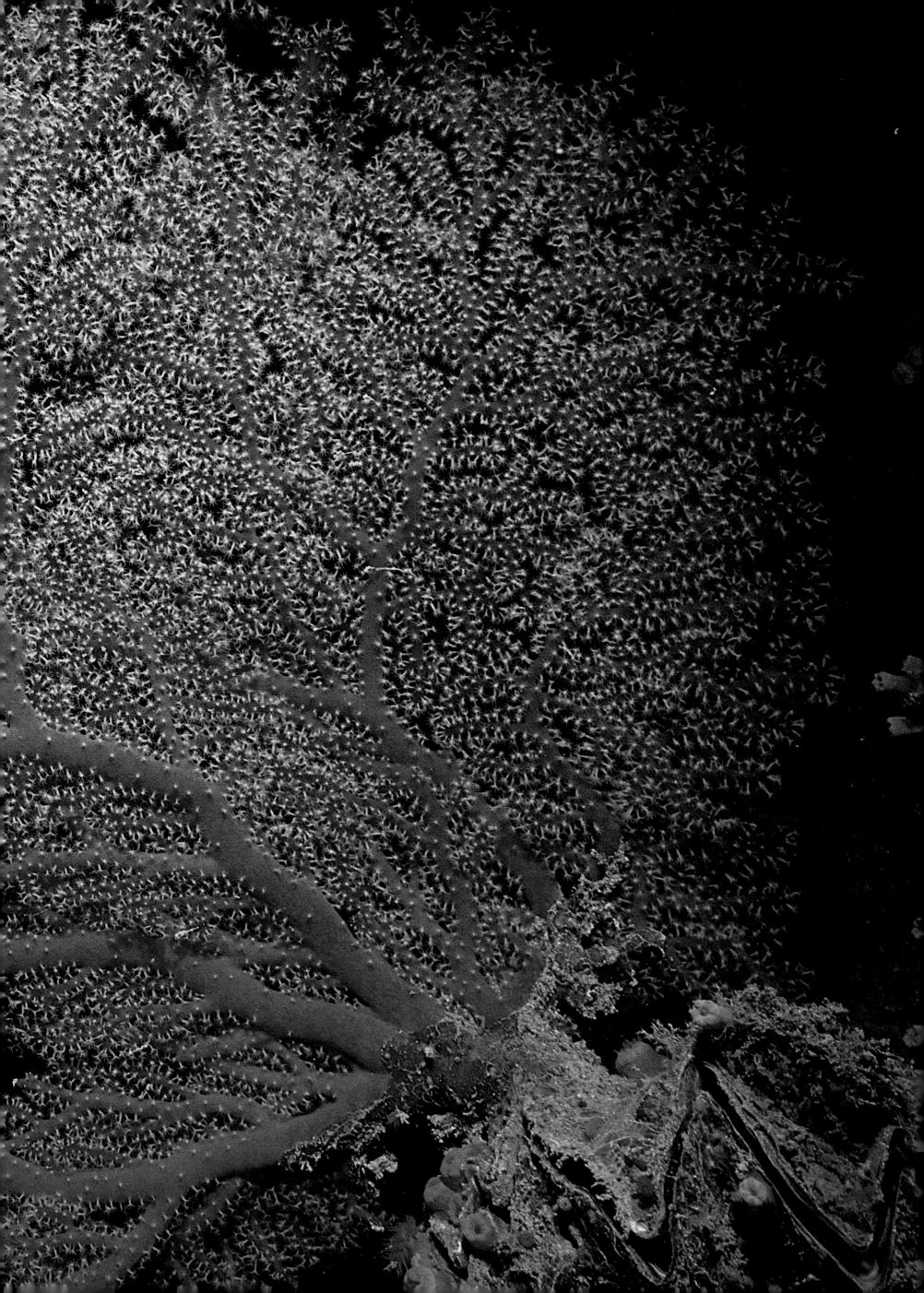

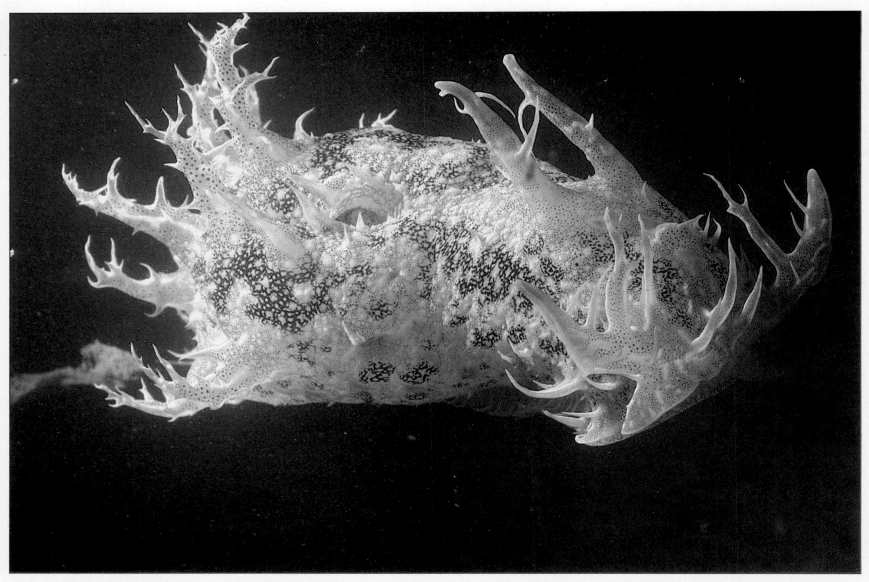

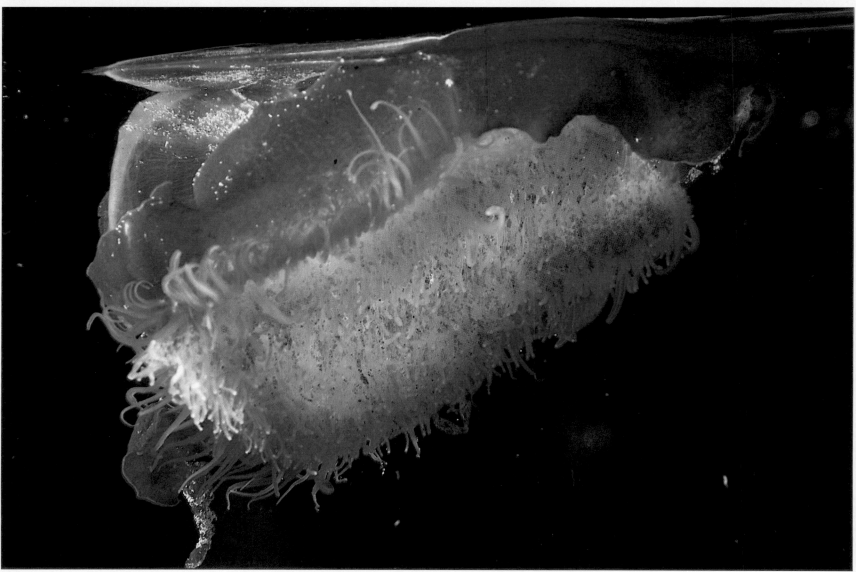

Best known of the fishes with tubular snouts is the sea horse (opposite), which swims upright and has a brood pouch in its tail. The sea hare (top) is related to nudibranchs but has an internal shell. By-the-wind-sailor is a jellyfish (bottom) with a ruglike fringe of tentacles and an off-center crest that catches winds for this oceanic drifter.

Rosy rockfish (top) and the tree fish (opposite top) are members of the rockfish family, a coastal temperate water animal sought by fishermen. Both the timid-looking blenny (bottom) and Coral Sea clownfish (opposite bottom) find protection among carpet anemones. No one knows why they are not stung and eaten. Sea fan corals look like flattened bushes but are animals working together for the good of the colony. This one is in the Caribbean (overleaf).

A close view of a surgeonfish (top) shows the yellow tail sheath where scalpel-sharp spines are hidden. Adult clownfishes do not travel far from the anemones they adopt (bottom). Both the sculpin (opposite top) and scorpionfish found off Florida (opposite bottom) belong to the worldwide scorpionfish family that includes the rockfish and stonefish.

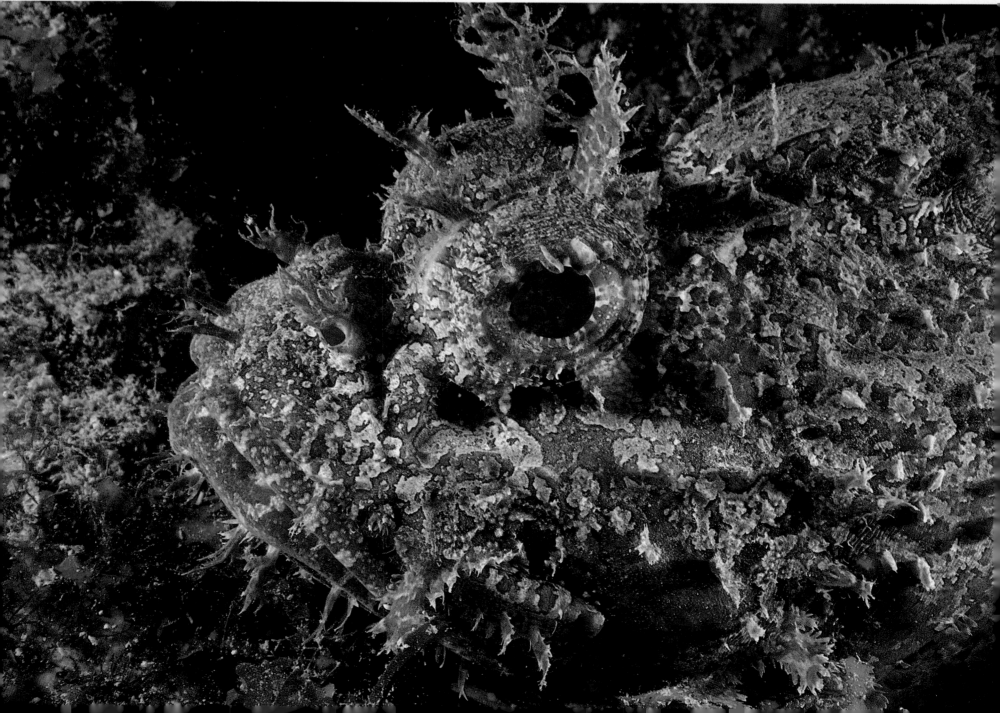

One of the scorpionfish clan's more dangerous members is the venomous lionfish, which is also known as the zebrafish. The normal swimming posture of outspread fins (above) *suggests it's small but feisty; a close-up* (opposite bottom) *defines the sharp spines that deliver the poison. It has been seen attacking and puncturing animals. The soapfish, a tropical Atlantic resident, has an irritating mucus on its skin that renders it unappealing to even a spiny sea star* (opposite top).

Caribbean fire corals extend their tiny tentacles to feed at night (top), as do soft tree corals, such as in Truk Lagoon in the North Pacific (opposite). The tree corals pump seawater through spicule-reinforced membranes to keep their shapes. Tealia anemones (bottom) are stalked animals that capture meals, day or night, by stunning prey with nematocyst-loaded tentacles.

A blood sea star checks gorgonia coral with a ray (**top**); *the aptly named pugnacious nudibranch* (**bottom**) *preys on fellow sea slugs; meanwhile, back in the South Pacific, another type of soft tree coral is feeding* (**opposite**). *At night, a basket star, which is related to brittle stars, spreads its pale arms to feed, just as the gorgonia under it is doing* (**overleaf**).

*A sheep crab (top), without the usual camouflage of plants or sponges, ambles across
the seabed. Red gorgonia coral (bottom) is widespread in the world ocean and has
been found at depths of 3,000 feet. The sand rose anemone (opposite top) seems to have
its mouth in the middle of a sunburst; a tiger grouper hides behind a coral head
(bottom) off Grand Turk Island east of Cuba.*

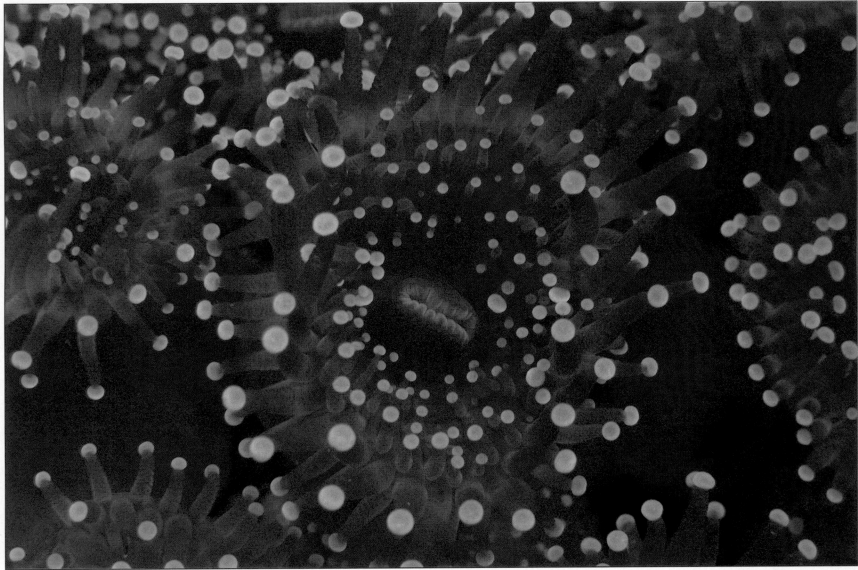

Reef corals, which are small algae-breeding colonies of animals, need clear, sun-lit water heated to at least 64 degrees F. to raise their edifice complex (top). *A closer look at a tropical reef shows diversity: yellow and red soft corals mix with sponges and hard corals* (opposite top). *Corynactis is a club anemone with varying colors* (bottom) *that is related to corals. A group of jelly animals widespread in the ocean's midwater zone are the long, delicate siphonophores* (bottom). *Orange cup corals hunker down in daylight when predatory grunts and snappers are active* (overleaf).

Undulations of the stingray's broadened pectoral fins (top) keep it moving; the gray angelfish (bottom) propels itself with both its tail and side fins. A tiny jeweled cleaning shrimp climbs anemone tentacles (top right), perhaps to service a fish. An orange ball anemone eats a small fish (bottom).

The gray whale, one of the few large whales to swim in shallow coastal waters where it acquires barnacles and eats tiny kelp animals, annually migrates 5,000 miles across the Pacific from Arctic seas to the lagoons of Baja California, Mexico.

*Unfazed by ice floes, a pod of white beluga whales starts a graceful dive. These are toothed whales, and dine on large and small fish. The larger, knobby-nosed humpback whales (**left** and **above**) lunge through schools of small fish with open jaws and trap prey with their mustachelike baleen fringe. Accidentally, this may include sea gulls more brash than wise.*

Killer whales, or orcas, are also toothed hunters with highly developed brains
and family relationships. They usually hunt fish together, but are bold
and fast enough to capture seals and sea lions—even at the surf line
(above, preceding pages). Orcas, like all whales, are air-breathing mammals.

Sea lions, seals, and walruses are flipper-footed sea mammals that have adapted to living on land and in the ocean. Tiny ears mark the California sea lion (opposite top) and shaggy Steller's sea lion (opposite bottom), while the harbor seal (top) and walrus have no external ears. Both male and female walruses have large canine teeth that grow into tusks (overleaf).

A harp seal mother and her creamy white pup vividly illustrate why sealers were interested in harvesting only the pup's pelt (above). Adult harp seals fish and frolic in the Arctic slush (opposite top). A mother hooded seal warns that she will defend her pup (opposite bottom).

Sea otters warm themselves with thick fur rather than blubber, and must eat often. They enjoy a wide diet but mostly consume urchins, squid, shellfish, and crabs. Of the starfish, only its innards are consumed. Laying on its back on the surface, sometimes wrapped in kelp, the Pacific sea otter often holds a stone on its chest to crack shells.

California sea lions could claim credit for originating body surfing.
Witness their skillful glide toward shore within the swelling
wavefront (top). A blond Australian sea lion keeps its large brown
eyes open under water (bottom).

A high-domed hogfish in the Sea of Cortez (top) wields powerful jaws that can crush most hard-shelled animals; a Pacific manta ray cups plankton to its mouth with modified pectoral fins (bottom); anchovies school for protection in the Sea of Cortez (opposite top). The Nassau grouper, which can change colors and scale patterns to match its surroundings, grows to weigh several hundred pounds (bottom).

Pilot whales usually travel in groups of 15 to 200, but this immature animal (top) off Hawaii's Kona coast seems to be alone. A school of needlefish swim sinuously on or near the surface of the open ocean (bottom). Egg-laying sponges spawn by releasing ova through their vents and canals (opposite), which may draw snappers or a multicolored wrasse looking for an easy meal.

*The green moray eel is checked out by a cleaner wrasse on a sponge-
encrusted reef (top). Feather duster tube worms spread their gills to gather
food as well as oxygen (bottom), while the Christmas tree worm (opposite)
channels food to its central stem, where it funnels into the worm's gut.*

When a group of coral polyps eats, the whole colony enjoys the meal (top); a spawning white gorgonia coral attracts an unusual concentration of Foureye butterflyfish (opposite top). A close-up of a spiny sea star arm reveals the animal's pincers and gill tissue around each spine (bottom); the white spot on the sunflower star's hub is a wound that may heal, or end killing the animal (opposite bottom).

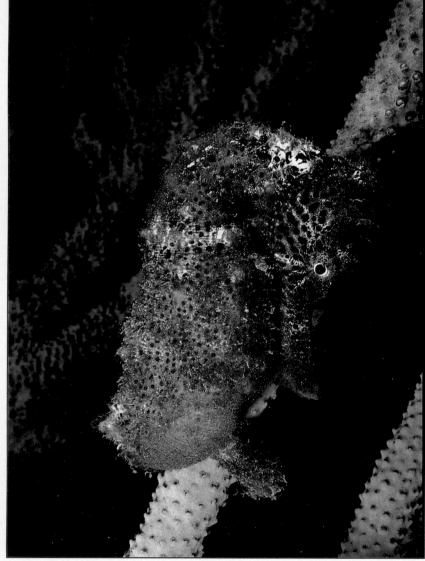

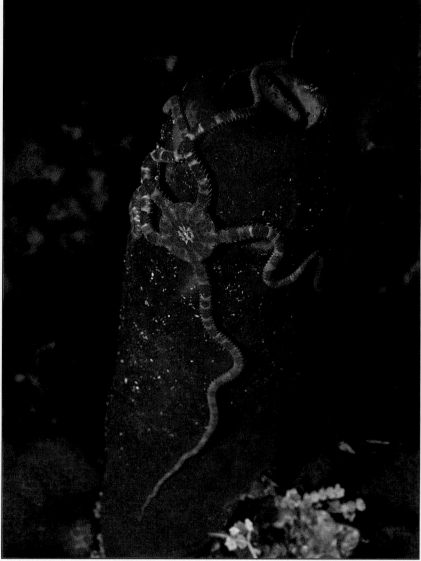

Crown-of-thorns starfish even go after gorgonia (top); a longsnout sea horse curls its prehensile tail around coral (bottom left). The brittle star, a fragile relative of the sea star, climbs up a sponge (bottom right) individual lightbulb tunicates, or sea squirts, reach for the Caribbean sun (opposite).

Young groupers, which are related to sea bass, take advantage of any protection offered by
sponges off Mexico (opposite top) and the Bahamas (bottom). These examples of the dorid
(top) and eolid (bottom) types of nudibranchs show how they breathe through gill-like tissues
on their backs. The dorid's is in bunches and the eolid's are arranged like fleshy spikes.

*A male sergeant major fish patrols a nest of purple eggs (top) lain by its
mate, while a spotted moray patiently waits for a wrasse to finish the
clean-up job (bottom). Opal sweepers school about a spray of feeding
black coral at night (opposite).*

Most of the turtles in the world ocean are threatened with extinction. A young green turtle (top) and a loggerhead turtle with suckerfishes upside down on its shell (bottom) swim along sandy shallows; the hawksbill's profile (opposite top) shows how this turtle got its name; Kemp's Ridley sea turtle is one of the endangered tropical species (opposite bottom).

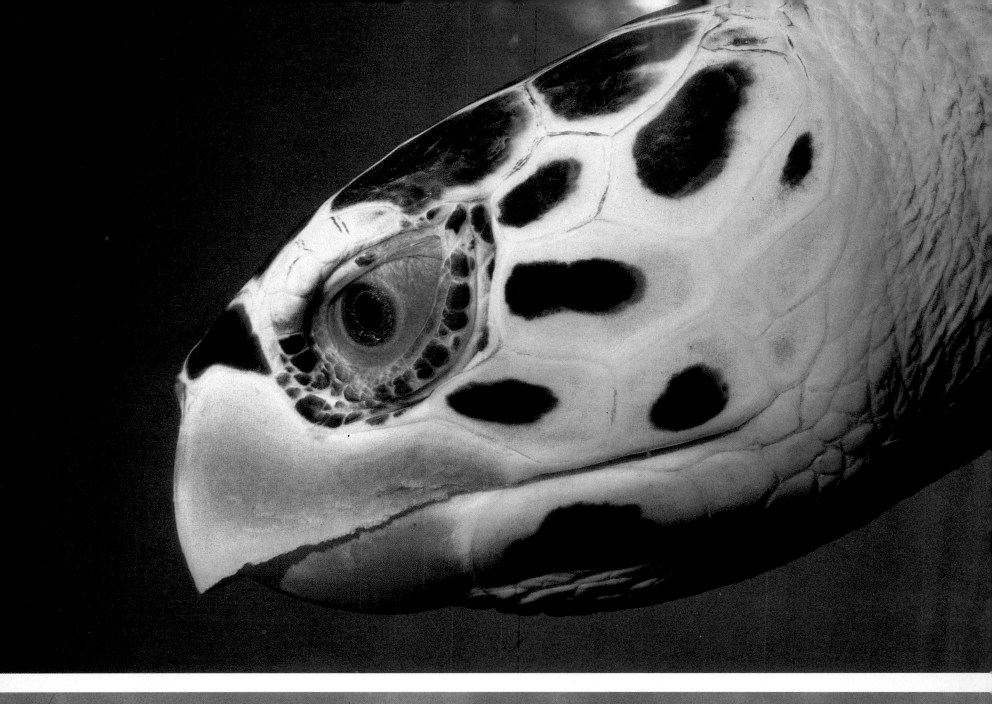

Chinstrap penguins hunt krill in the Atlantic side of Antarctica and enjoy resting on a handy iceberg (preceding pages); *a mother chinstrap and her chick* (above) *show how appropriate the name is for adults. Adélie penguins* (opposite) *actually live on the ice continent, enduring below-freezing winds and building nests of pebbles. Fast as seals in the water, Adélies waddling over the ice suggest chronic, and comic, awkwardness* (overleaf).

King penguins regally ignore a perplexed-looking young elephant seal (bottom), though adult bulls are decisive and vocal (top) if they think a rival is near. The mother harbor, or common, seal is watchful while her pup depends on her (opposite top), but soon the young ones will have to fend for themselves under water. Chinstrap and a few of the related Adélie penguins trek over Antarctic ice (overleaf), probably to where they can reenter the sea.

A common southern sea bird is the blue-eyed shag, also known as the king cormorant (opposite top); the Arctic counterpart of the penguin is the puffin, seen above with its beak in full breeding colors. A Weddell seal pup at a breathing hole (opposite bottom) is another resident of the Antarctic coast.

Leopard seals (top and bottom) *are major natural predators of the speedy Antarctic penguins, making most of their kills in waters where ice hampers the birds' exit. Weddell seals* (opposite) *live in one of Antarctica's coldest spots, where the frigid Weddell Sea pushes cold-water masses toward the Equator. Pups weigh around 60 pounds at birth and start to shed their gray coats at two weeks.*

Steller's sea lions (top and bottom) *and California sea lions* (opposite)
*have ears and share many physical characteristics. However, Steller's
bull tips the scales at better than half a ton, outweighing its
California cousin by far.*

Polar bears are superbly adapted to the Arctic halfway house where ice and sea meet.
Studies of tranquilized bears show that their hairs collect the weak polar sunlight,
converting 95 percent to heat, little of which escapes. They are also excellent swimmers,
capable of crossing miles of sea and slush with the aid of webbed paws.

Red-orange finger sponges grow on a reef off Belize (above). The sheepshead's strong teeth and jaws enable it to crush hard coral cups to reach the polyps (opposite top). At a cleaning station a tiger grouper opens wide for neon gobies, while a juvenile Spanish hogfish tidies around the tiger's gills (bottom).

When red gorgonia coral feed (opposite), *little looks like it can escape.*
A type of bluebell tunicate filters sea water through an internal
basketlike sieve that efficiently extracts organic particles smaller than
those caught by the coral (above).

Fire corals look like stubby trees instead of living animals **(top)**. *Dorid nudibranchs* **(bottom)**, *often use irritating chemicals to ward off predators. Soft tree corals, like the one atop the giant clam* **(opposite top)**, *overpower hard corals with other agents, though they may not stop fish like these opal sweepers. Yellow and black rock beauties gobble a nest of eggs after driving away the defending sergeant major fish* **(opposite bottom)**.

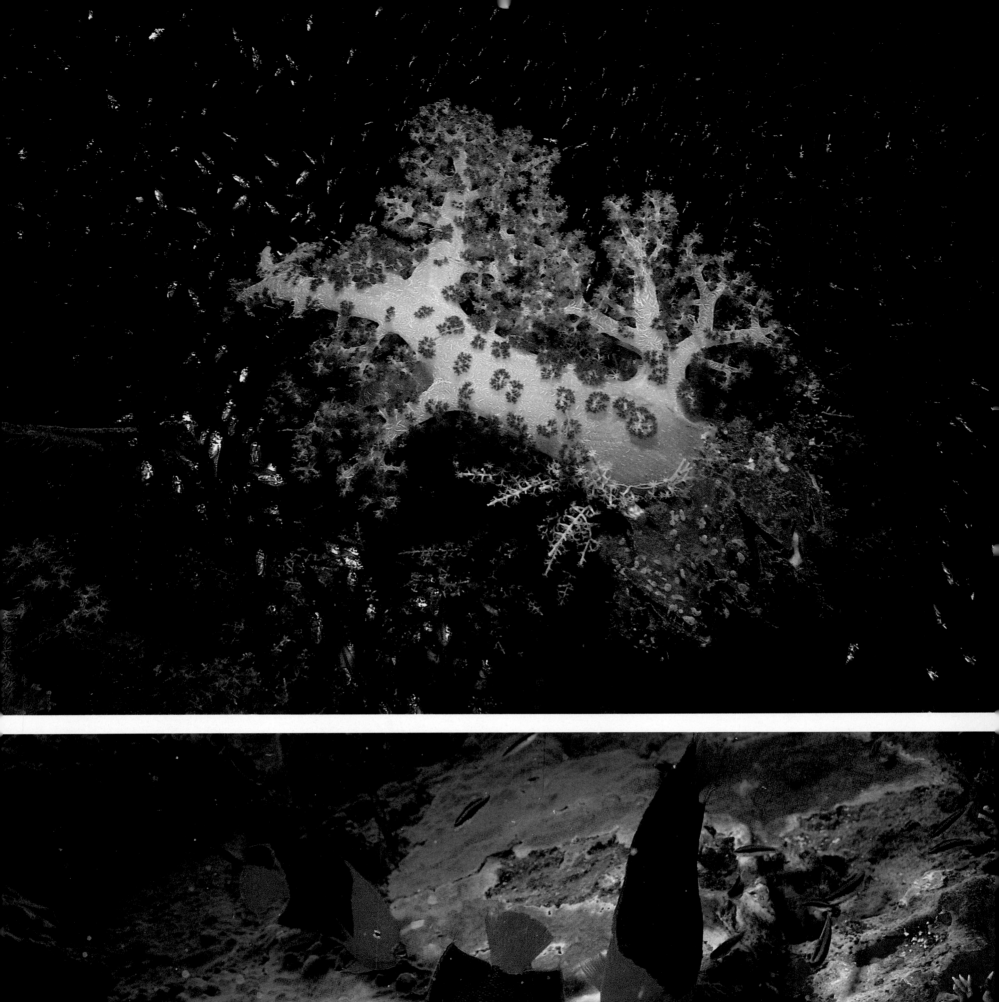

*A close-up reveals brain coral polyps searching for food (top); the
California arminid nudibranch eats polyps and suggests how diverse
are sea slugs' forms (bottom); blue damselfish hang above feeding leather
coral like one of nature's mobiles (opposite).*

INDEX BY PHOTOGRAPHER